Kuhn's *The Structure of Scientific Revolutions* Revisited

Routledge Studies in the Philosophy of Science

Kuhn's *The Structure of Scientific Revolutions* Revisited

**Edited by Vasso Kindi
and Theodore Arabatzis**

Routledge
Taylor & Francis Group

NEW YORK   LONDON

First published 2012
by Routledge
711 Third Avenue, New York, NY 10017

Simultaneously published in the UK
by Routledge
2 Park Square, Milton Park, Abingdon, Oxfordshire OX14 4RN

First issued in paperback 2014

Routledge is an imprint of the Taylor and Francis Group, an informa business

Library of Congress Cataloging-in-Publication Data
    Kuhn's The structure of scientific revolutions revisited / edited by Vasso Kindi and Theodore Arabatzis.
        p. cm. — (Routledge studies in the philosophy of science ; 12)
    Chiefly proceedings of a conference held in Aug. 2008 in Athens, Greece.
Includes bibliographical references (p.) and index.
    1. Kuhn, Thomas S. Structure of scientific revolutions—
Congresses.   2. Science—Philosophy—Congresses.   3. Paradigm
(Theory of knowledge)—Congresses.   I. Kindi, Vasso,
1957–   II. Arabatzis, Theodore, 1965–
    Q175.K953K84 2012
    501—dc23
    2012021603

ISBN 978-0-415-80855-2 (hbk)
ISBN 978-1-138-91087-4 (pbk)
ISBN 978-0-203-10315-9 (ebk)

Typeset in Sabon
by IBT Global.

Contents

PART III
Implications

Figures and Tables

FIGURES

TABLES

Acknowledgments

We would like to thank the contributors to this volume for trusting us with their work and for their enthusiasm for the project. We are also grateful to all the participants in the international conference in which part of this book was first conceived. The conference on "T. S. Kuhn's *The Structure of Scientific Revolutions*: Impact, Relevance and Open Issues" took place in Athens, Greece, in August 2008 and was made possible by the support of the J. F. Costopoulos Foundation and the Eugenides Foundation.

We would like to acknowledge the support of the University of Athens and our colleagues, most importantly among them Kostas Gavroglu, Aristides Baltas, Aris Koutoungos, and the late Pantelis Nicolacopoulos, whose efforts, beginning in the 1980s, have led to the creation of an academic community of philosophers and historians of science in Greece. In that process Kuhn's book was instrumental and contributed significantly to a change of the Greek philosophical and historical landscape, by attracting to the study of philosophy and history of science scholars with a background in the natural sciences, who brought into philosophy some of the rigor found in their home disciplines.

Excerpts from Kuhn's *The Structure of Scientific Revolutions* are reprinted with permission of the University of Chicago Press (© 1962, 1970 by The University of Chicago. All rights reserved. Published 1962. Second Edition, enlarged, 1970). Last but not least, we would like to express our thanks to our editors, Felisa Salvago-Keyes and Catherine Tung, for their enthusiasm about our project and their unfailing support; to John Rogers for his meticulous editing of the manuscript; to Michael Watters for managing the production of the manuscript with efficiency and care; to Antonia Pavli for preparing the index; and to three anonymous referees for their encouragement, helpful suggestions, and constructive criticism..

1 Introduction

Vasso Kindi and Theodore Arabatzis

Anniversaries mark the time spanned and, by commanding pause, they occasion retrospection and renewed assessment of the events celebrated. In the present volume we revisit Thomas S. Kuhn's *The Structure of Scientific Revolutions* (SSR) to reassess, fifty years after its first publication, its value, impact, and current relevance.

The book has been characterized as, and we believe it is, a modern classic. It has influenced directly and indirectly a hugely wide range of disciplines, fields of inquiry, even areas of popular culture. Terms such as *paradigm* and *paradigm shift, incommensurability,* and *normal science* have found their way nearly everywhere: in academic journals and books of the natural and social sciences, of the humanities, of engineering and of medical science, in theological texts, literary essays, speeches of politicians, journalists' reports, cartoons and comic strips, even on a San Diego charter boat which bears the name "Paradigm Shift" (Rowbottom 2011, 112). In the wake of the book's publication, disciplines were shaped or reformed (for instance, STS and history of science) and the dominant approach in philosophy of science (i.e., logical positivism) receded into the background.

SSR's original reception was not, however, thunderous, as one might expect given how provocative the book was eventually registered as being. It was well received at first by natural scientists, as Kuhn was always eager to point out, while historians of science and philosophers, for the most part, expressed initially minor reservations and doubts. The scientists recognized aspects of their practice in its pages while the philosophers' and historians' attention was arrested by the unusual injection of historical material into the book. Before Kuhn's SSR, historical examples were used, if ever, in philosophy of science texts merely for purposes of illustration. Philosophers would advance a priori considerations regarding science and they would occasionally resort to some historical example to illustrate and lend auxiliary support to what they said. It was taken for granted, by both philosophers and historians, that history of science solidly confirms the standard conception of scientific rationality and progress. The idea was that scientists follow the scientific method, which involved the confirmation and falsification of statements, ensuring thus the steady expansion and accumulation of scientific knowledge. In Kuhn's book, however, history was not a repository of events called upon circumstantially from the background, but a major player in shaping his model of science. The exact way Kuhn made use of history in SSR is still debated, but it is clear that he blended internal and external factors in his analysis of the scientific phenomenon and that the end result escaped standard disciplinary classifications. SSR was an original composition, bringing together historical, philosophical, sociological, and psychological considerations. This fresh approach can explain the book's original appeal, the early questions and reservations, but also the subsequent vehement critique which burgeoned mostly in philosophy. It was feared that the injection of nonepistemic parameters into the study of science would expose scientific knowledge to the ravages of contingency and irrationality. Kuhn's notions of revolution and of incommensurability invited arguments which charged him of rampant relativism and idealism and of introducing personal idiosyncrasies and mob psychology into science proper.

Kuhn tried to explain, to clarify, to somewhat modify, and to respond, but never did he renounce his stance in order to appease his critics or indulge his followers. He stood steadfastly by the thesis of incommensurability and that of discontinuity and radical change, trying at the same time to secure the rationality of science and a place for the world in his schema. Some of his most fervent advocates drove his views to the extreme, claiming the construction of reality by power politics, but he, firmly, but not unworriedly, resisted idealist and relativist interpretations. He dismissed them as absurd but he simultaneously feared that inescapable initial observations from the history and sociology of science may lead, by a slippery slope, to quite unacceptable conclusions in that they would deprive science of its authority. Concerns over the authority of science, which was thought to be corroded and threatened by Kuhn's book, fueled the criticism against it and his work was increasingly marginalized and dismissed as unenlightening, muddled, unfounded, or plainly wrong. However, as the hostility in academic circles mounted, mostly in the quarters of philosophy, his ideas quietly and rapidly spread to all areas of thought and became so familiar to the point that they were not considered worth discussing.[1] Many articles dwelling on his work found their way into relevant academic journals, but gradually the tide of philosophy of science flowed toward more traditional epistemological and metaphysical issues (for instance, induction, confirmation, explanation, and realism). History of science also moved in directions that were not, at least avowedly, attributed to Kuhn. Historians left his work behind as still tied to a by-now suspect internalist conception of science (which, ironically, Kuhn helped to bring down) and as belonging to the outmoded genre of grand narratives. In general, Kuhn's work was seen as having been tried and convicted (of irrationality, relativism, idealism, dogmatism) and it was claimed that now it was time to move on. His ideas, though, were not eclipsed and congealed into the background against which the new developments in philosophy and history of science were unfolding. Previous certainties, which formed the context in which SSR made its appearance, slowly gave way to a new understanding of science ushered in by this book; its insights, which were once considered provocative and controversial, eventually were treated as pallid platitudes. For instance, after Kuhn's SSR, even though the credit was not openly given to him and the other figures of the historical turn in philosophy of science, few would assent to the triumphalist idea of science, to the systematically cumulative growth of knowledge, to anachronistic readings of the past, to fixed and unadulterated criteria of rationality, to a clear-cut theory-observation dichotomy, to unproblematic reductions and translations of theories, to identifying science with theories and sets of propositions. So, despite the marginalization/overshadowing of Kuhn's book by subsequent developments, it had a strong, even if latent, contribution to the change of landscape regarding the study of science.

We do not mean to imply that SSR was not extensively discussed. Many of its themes were taken up in numerous articles written in the 60s, 70s, and 80s. Rather, the point we want to make is that the book's philosophical reception was shaped, for the most part, by the debates and climate in philosophy of science in the 1960s and 1970s. It was heavily criticized for not meeting the standards of the then-dominant view: it was found to be rather vague, inconsistent, not technical enough, philosophically naïve, and open to charges of promoting relativism and irrationalism, always by the yardstick of a philosophy which was itself one of the targets of the book's author. On the other hand, most of its defenders, who often came from sociology of science, did not challenge the framework of debate and celebrated some of the same points that the book's critics took to be its defects, without concentrating on the actual theses developed in it. The controversy regarding Kuhn's work rehearsed more or less the very same arguments for more than forty years. What we often had was the invocation of standard topoi irrespective of what Kuhn actually said. Kuhn's nontechnical and unassuming language and the description of experiences and phenomena familiar to anyone who is even remotely acquainted with science also discouraged a more thorough engagement with the book. The deceptively simple exposition invited the complacent attitude that it is a book very easy to understand and there is no point in probing further into what it has to offer. The result is that its innovative ideas and their implications have not been fully appreciated.

The present volume originated in an international conference which took place in Athens, Greece, in August 2008. It includes papers presented at the conference as well as contributions from other Kuhn scholars. The aim of the volume is to take a fresh look at *The Structure of Scientific Revolutions*. We concentrate on particular issues addressed or raised in the book in light of recent scholarship and without the pressure of the immediate concerns scholars had at the time of the book's publication. In recent years there has been novel research on a number of topics relevant to SSR's concerns, such as the nature and function of concepts, the complexity of logical positivism and its legacy, and the relation between history and philosophy of science. Furthermore, there has been extensive research on the historical episodes in the development of science which were discussed in SSR. The scholars who contribute to this volume have all been actively involved in this kind of philosophical and historical research. They have long worked and published on Kuhn and/or the problems that his work posed. Thus, they bring relevant results to bear on SSR's topics, but here they are looking at the book from novel perspectives, trying to assess its contribution and reconsider the debates it gave rise to.

The chapters both explicate and assess. They make new connections and undertake new approaches in an effort to understand the book's legacy and significance and to come to terms with problems that have mired scholars in the previous decades. They reflect recent trends in philosophy of science: a systematic engagement with the history of philosophy of science (HOPOS), a turn to the nitty-gritty details of scientific practice and an associated naturalistic bend, and a tendency to "lower the tone" (Shapin 2010), that is, to demystify the grand issues of rationality and realism. Thus, our contributors offer sensitive and well-disposed (albeit often critical) readings of Kuhn's philosophy of science, situate it within the past and contemporary philosophical scene, and examine its implications for current issues.

Most of the HOPOS literature on twentieth-century philosophy of science has focused on the history of logical empiricism. That was, of course, badly needed, given the simplistic and often inaccurate "everyday image" of that philosophical movement (cf. Richardson 2007). Kuhn's work, on the other hand, has not received the historical attention it deserves. While SSR has generated a significant amount of scholarly work, very little of it is properly historical, that is, aiming at understanding Kuhn's work in its intellectual, social, and cultural context.[2] The chapters in the first part (*Origins and Early Reception*), by Gürol Irzik, James Marcum, and Rupert Read and Wes Sharrock, represent steps in this direction. They discuss the origins and early reception of SSR, situate the book in its context, and consider its targets and its legacy.

Irzik examines the relationship between Kuhn's work and the dominant philosophy of science at that time, an outgrowth of logical positivism (LP). Irzik admits that Kuhn attacked certain doctrines associated with LP and Karl Popper: the theory-observation distinction (probably unaware that Hans Reichenbach and Popper had already questioned it) and the confirmationist/falsificationist views of scientific practice. However, as Irzik argues, the "positivism" that Kuhn criticized is not primarily that of LP, but rather the "spontaneous philosophy" of older historians of science and textbook writers. The main tenet of that philosophy was the cumulative character of the development of science.

In tune with recent HOPOS scholarship, Irzik suggests further that the distance between Kuhn and the most important representatives of LP

(Reichenbach, Otto Neurath, and Rudolf Carnap) is smaller than we previously thought or, at any rate, than Kuhn portrayed it to be. According to Irzik, our understanding of the history of LP was shaped by Kuhn and its other critics (most notably, Paul Feyerabend and W. V. Quine). Kuhn's work, in particular, was for the most part responsible for the "everyday image" of LP. Recent HOPOS scholarship on the origins and development of LP has revealed the poverty of that image.

Marcum discusses the evolution of Kuhn's views during the 1960s. In particular, he provides an account of the road from the original edition of SSR in 1962 to the Postscript to the second edition in 1970, focusing on three steps along that road: the 1965 London colloquium on philosophy of science, Kuhn's 1967 lecture at Swarthmore, and the 1969 Urbana symposium on the structure of scientific theories. As Marcum shows, it was in that context that Kuhn refined his notion of "paradigm." Kuhn's distinction between two components of paradigm—"exemplar" and "disciplinary matrix"—emerged as a result of his attempt to respond to the criticism that he had received on those occasions. In the Swarthmore lecture, for instance, Kuhn suggested a narrower sense of "paradigm" as a concrete problem solution and claimed that paradigms, in that sense, function as mediators between scientific theories and the world.

Furthermore, Marcum traces the critical reception of Kuhn's distinction between normal and revolutionary science and examines Kuhn's efforts to draw that distinction in a satisfactory manner. At the end, Kuhn's solution to that problem was "sociological": to identify a paradigm one needs to identify, using standard sociological techniques (e.g., citation patterns), the corresponding scientific community. Marcum's chapter sets the stage for the subsequent discussion of paradigms and exemplars by Vasso Kindi and Thomas Nickles.

Read and Sharrock present SSR as a theoretical treatise in historiography and stress the historiographical fruitfulness of the image of science developed by Kuhn. They note Kuhn's turn from normative to descriptive/interpretive issues and, in particular, to a descriptive account of rationality. One of Kuhn's main historiographical points was that in order to understand adequately past scientific practice we should historicize rationality. For historiographical (if not for philosophical) purposes, contemporary criteria of scientific rationality should not be considered universally applicable.

Read and Sharrock read Kuhn in parallel with Peter Winch, who around the same time as Kuhn put forward similar arguments in the philosophy of social science. Both Kuhn and Winch argued that in order to understand others we need to contextualize their beliefs and actions, that is, to recover the intellectual, social, cultural, etc. contexts that endow those beliefs and actions with meaning.

Kuhn and Winch gave priority to the actors' self-understanding. To understand a practice we have to understand it from within. We have to make sure that current knowledge does not get in the way of understanding past knowledge. The right model here would be that of an anthropologist trying to interpret an alien culture. The problem with past science is that it often looks deceptively similar to contemporary science. Thus, the first step toward historical understanding is defamiliarization.

Read and Sharrock point out that Kuhn's (and Winch's) interpretive suggestions were misconstrued as gestures toward relativism. In the 1960s and 1970s most readers of SSR (especially in philosophy) were captivated by its prima facie relativistic and antirealist tone and bypassed very significant (perhaps the most significant) aspects of the book.

The chapters in the second part (*Key Concepts*), by Vasso Kindi, Thomas Nickles, Hasok Chang, and Jouni-Matti Kuukkanen, exemplify the turn to practice that has characterized post-Kuhnian history and philosophy of science. The focus of practice-oriented scholarship is on the epistemic process itself, as opposed to the products of that process. Kuhn's work was largely responsible for that practical turn and some of the novel concepts he introduced in SSR (e.g., paradigm and normal science) aimed at capturing salient aspects of scientific practice. The contributors in this second part revisit those concepts and metaphors: paradigm, normal science, scientific revolution, incommensurability, and "Darwinian evolution." They explain their role in Kuhn's philosophy of science and evaluate their implications for central philosophical issues, such as scientific rationality and scientific realism.

Paradigm, in the sense of exemplar, was emblematic of Kuhn's turn to practice. In her chapter, Kindi argues for an expansive understanding of "paradigm," encompassing "exemplar" and "disciplinary matrix," by suggesting a reading of exemplars inspired by Wittgenstein's analysis of rule following. Pace many commentators (including Kuhn himself), exemplars should not be sharply juxtaposed to rules. They go hand in hand: exemplars follow set rules, rules need exemplars to be correctly followed, and together they give rise to a tradition. Thus, there is a close link between the two main aspects of the paradigm concept, exemplar and tradition, and the notorious polysemy of the term is shown not to be problematic.

In the first part of the chapter, Kindi examines how Kuhn uses "paradigm" in SSR and in his subsequent writings and shows the problems that emerge from the attempts made to disambiguate the term and retain one or the other option. In the second part, she compares Kuhn's understanding of "paradigm" to Wittgenstein's homonymous term and proceeds to discuss how a paradigm, exemplar, or model is to be followed. She argues that the way an exemplar, functioning as model, is being copied and followed, setting a rule, is not dictated by the exemplar *qua* physical object but has to be learned in training and picked up in practice. The exemplar is already a symbol and the similarities between it and the subsequent instances of its application are being imposed and learned rather than discovered. In this sense the whole discussion of exemplars and rules by Kuhn has a logical rather than naturalistic character. He is not interested in studying human reasoning from the point of view of cognitive science but rather in presenting the grammar of the relevant concepts, that is, how they relate to each other and to other notions, how they make possible scientific practice. In connection to this she also considers the differences between the role of cases in case-based disciplines, such as medicine or law, and the way Kuhn understands the function of exemplars in scientific practice.

Exemplars and practice are also the focus of Nickles' chapter. Departing from the older literature on SSR, which was critical of a sharp distinction between normal and revolutionary science, Nickles reexamines this distinction by focusing on the origin and trajectory of exemplars. He argues that the birth of new exemplars is often the outcome of "normal" research. After their birth they evolve within the bounds of normal science. Nickles stresses the dynamic character of normal science and challenges Kuhn's account of it as a static, monolithic enterprise. Normal science isn't as cumulative as Kuhn portrays it to be.

Furthermore, some exemplars persist, suitably reconstructed, across revolutionary breaks. A focus on exemplar-based practices makes it possible to recover some elements of continuity across revolutionary breaks. Scientific revolutions are less radical than presented in Kuhn's model of scientific development.

Thus, the evolution of exemplars within a paradigm and their persistence across paradigm change narrow the gap between normal and revolutionary science. The former, like the latter, is a developing enterprise, and the latter, like the former, preserves previous practices.

The turn to practice has also led Kuhn to a reappraisal of the so-called context of justification. As Nickles notes, in scientific practice, normal and revolutionary alike, justification is forward looking. From the point of view of the practicing scientist, the future potential of a scientific achievement is its most significant characteristic.

Nickles hints at Kuhn's evolutionary account of scientific development, which is characterized by "variation," "selection," and "transmission." This evolutionary analogy is fully taken up in Kuukkanen's chapter. He points out two aspects of the analogy in question: a diachronic aspect (emphasized in SSR) and a synchronic aspect (developed in Kuhn's later work). The former is associated with a rejection of teleology and an emphasis on the survival of the fittest paradigm. The latter is associated with "speciation," an increasing "specialization," and a "proliferation of scientific specialties."

By reading Kuhn's philosophy of science through an evolutionary lens, Kuukkanen is led to some of the theses also defended by Nickles. First, paradigms turn out to be evolving entities, changing gradually by the collective action of scientific communities. Revolutions lose their global, earth-shattering character and are transformed into local and, largely, continuous processes. Parts of the older paradigms (or lexica, in Kuhn's later terminology) are retained across revolutionary change. Second, incommensurability continues to play a crucial role in the development of knowledge but now acquires a significant cognitive function. Far from being a threat to the rationality of scientific development, it is seen as a primary motor of progress. Incommensurability, now understood as the impossibility of context-independent translatability, hampers communication, leads to the linguistic isolation of a scientific community, and thus allows the autonomous development of a specialty.

This sympathetic account of incommensurability is in stark contrast with the negative and hostile reception of that notion during the early reception of Kuhn's work. Nevertheless, some of the implications of incommensurability, duly pointed out in the older literature on SSR, remain in place. Most notably, incommensurability still belies "convergent realism." Even though scientific practices are adapted to the world, they also shape it, conceptually and materially.

A similar, positive stance toward incommensurability pervades the chapter by Chang. This chapter employs incommensurability as a tool for probing one of the main historical episodes treated in SSR, the Chemical Revolution. Chang finds useful the Kuhnian framework for interpreting that episode, one of whose main aspects was the transition from a phlogiston-based to an oxygen-based chemical paradigm.

Following Kuhn's account in SSR, Chang distinguishes different senses of incommensurability: with respect to concepts, methods, epistemic values, and problems. He then argues that the Chemical Revolution was characterized by incommensurability in methods, values, and problems, but only to a minor extent by conceptual incommensurability. As a result of these multiple dimensions of incommensurability, the choice between the phlogiston-based and the oxygen-based paradigms was underdetermined. One can thus understand why that chemical controversy was difficult to resolve and why several chemists adhered to phlogiston well after the Chemical Revolution had been over. In all, Chang demonstrates the pragmatic historiographical value of incommensurability, which was seen by Kuhn as his main philosophical innovation.

At the same time, Chang stresses the substantial commensurability between the phlogiston-based and the oxygen-based paradigms. At the "operational" and "phenomenal" levels, the meaning of chemical practices remained remarkably stable and that stability was a prerequisite of the conceptual incommensurability that arose at the "theoretical" level. Chang's motto is "back to *Structure*," that is, back to the rich and multilayered characterization of incommensurability that Kuhn originally put forward, and, by implication, away from the narrow linguistic conception of incommensurability that Kuhn favored in his later years.

The chapters in the final part (*Implications*), by Hanne Andersen, Alexander Bird, and Alan Richardson, discuss the continuing relevance of SSR and its implications for current issues and debates, namely the problem of conceptual change, the naturalistic turn in philosophy of science, the merits of cognitive and sociological approaches to science, and the significance of philosophy of science for historiography.

Andersen's chapter examines how Kuhn's views on conceptual change evolved after SSR. Starting from Kuhn's early advocacy of the family-resemblance account of concepts, she moves to the taxonomic approach to concepts and incommensurability that Kuhn developed in his later years. Andersen exhibits the elements of continuity between Kuhn's early and later work and stresses some of the themes that we also encounter in the other chapters: Kuhn's debt to Wittgenstein as regards the importance of family resemblances for the meaning of concepts (Kindi), Kuhn's evolutionary account of scientific change (Kuukkanen), and Kuhn's reinterpretation of "gestalt switches" from an actors' category to an analysts' category (Chang).

Furthermore, Andersen explores the connections between Kuhn's account of concepts and related research in cognitive science and notes that Kuhn had very little to say about the process that leads to conceptual change. This gap in the Kuhnian scheme can be filled by cognitive history and philosophy of science (HPS), which may explain, among other things, the crucial role of problem situations, exemplars, and anomalies in concept transmission and change (cf. Andersen, Barker, and Chen 2006).

Finally, Andersen points out how Kuhn's account can be developed in order to meet the philosophical challenges posed by interdisciplinary research. She argues that in order to understand how interdisciplinary communication and collaboration are possible, one has to focus on the overlaps and interconnections of incommensurable taxonomies.

The following chapter, by Alexander Bird, discusses further cognitive HPS. Bird considers Kuhn's naturalistic approach one of the most pioneering and significant characteristics of his philosophy of science. Kuhn's approach was naturalistic in two respects: first, he used the history and sociology of scientific practice as a source of philosophical insights; second, he drew upon the findings of cognitive psychologists to support philosophical theses (e.g., about the theory-ladenness of scientific observations).

Both of these naturalistic elements have been further developed by the sociology of scientific knowledge (SSK) and by cognitive HPS respectively. Bird sees only the latter as the "true heir" of Kuhn's legacy, for having come to terms with Kuhn's "internalism." Notwithstanding Kuhn's emphasis on sociological factors, Bird argues persistently that Kuhn was an internalist, that is, he regarded science as driven primarily by its own internal dynamic. External factors enter the Kuhnian model either in the very early stages of the formation of a paradigm or "in determining the timing of a scientific development" (212). SSK, on the other hand, has adopted "a predominantly externalist approach" (211). It is, therefore, against the spirit of Kuhnian philosophy of science.

Bird's interpretation of SSK as an externalist approach to science might be controversial for two reasons: first, several SSK-inspired studies (e.g., Pickering 1984) have focused on microsocial factors that Bird considers internal to science; and, second, some scholars sympathetic to SSK have called into question the very significance of the internal/external dichotomy.[3] Be that as it may, Bird's portrayal of Kuhn as an internalist is very plausible and would occasion no surprise to those familiar with Kuhn's historical work, especially with his *Black-Body Theory and the Quantum Discontinuity*, which provides a deep, but exclusively internalist, analysis of the origins of quantum theory (Kuhn 1987).

In closing, Bird suggests the need for an integration of the historical-sociological and the cognitive dimensions of Kuhn's naturalistic approach. That integration could have exemplars as its vehicle. A historical-sociological understanding of the role of exemplars in scientific practice could only be achieved by considering their cognitive-psychological function.

The final contribution to the volume, by Alan Richardson, has as its point of departure the first chapter of SSR. In that chapter Kuhn argued that the image of science as a cumulative enterprise practiced by heroic individuals led to banal or sterile historical questions, concerning, for instance, the precise timing of scientific discoveries and the allocation of credit for them to particular scientists. Kuhn suggested the need for a new image of science, which would, in turn, give rise to new, more interesting and fruitful historical questions.

Richardson draws inspiration from Kuhn's philosophical historiography of science for renewing the historiography of philosophy. He subjects the historiography of philosophy to a searching critique along the lines of Kuhn's criticism of "positivist" historiography of science. The former, just like the latter, leads to "unanswerable, misposed, or uninteresting" questions (237) and presents philosophical activity as dealing with the same "perennial" problems throughout its history. This unhistorical image of philosophy is achieved through a misleading interpretive strategy: only certain parts (the "philosophical" parts) of the work of past philosophers are read, and they are interpreted without any effort to contextualize them, either within the wider corpus of a philosopher's work or within the intellectual and cultural context of his time. Not surprisingly, many historical facts about the philosophical past have difficulty fitting within this unhistorical straitjacket.

Richardson, in contrast to Bird, maintains that Kuhn undermined the image of science as an autonomous enterprise and argues that a historiography of philosophy inspired by Kuhn's work would similarly have to pay proper attention to the cultural context of philosophical activity. A renewal of the historiography of philosophy might even alter philosophical practice itself by transforming the "self-understanding of the working philosopher and yield[ing] a richer, more satisfying set of philosophical projects" (248).

All in all, we hope that this volume will revive the interest in SSR, which, with a few exceptions (e.g., Wray 2012), has not recently enjoyed the focused attention of scholars.[4] Reading the chapters one is struck by the freshness and continuing relevance of Kuhn's work. Kuhn was a very acute observer of scientific life and he brought his historical expertise to bear on his philosophical interpretation of science. We believe that the picture of

Kuhn that emerges in the following contributions does justice to his powerful and subtle philosophical account of scientific development.

NOTES

1. Cf. Hacking (1990, 355): "Kuhn is too well known to need discussion."
2. A notable exception is, perhaps, Fuller (2000). This work, however, is overtly hostile to Kuhn and this attitude prejudices the contextualization of his work. Cf. also Friedman (2008), Kindi (2010), and Wray (2011a).
3. For an enlightening discussion, see Shapin (1992).
4. Most of the recent book-length studies on Kuhn discuss his philosophy of science in general, without focusing on SSR. See Horwich (1993), Hoyningen-Huene (1993), Fuller (2000), Andersen (2001), Bird (2001), Nickles (2002), Sharrock and Read (2002), Marcum (2005), Gattei (2008), Preston (2008), Wray (2011b).

REFERENCES

Andersen, Hanne. 2001. *On Kuhn*. Belmont, CA: Wadsworth Publishers.

Andersen, Hanne, Peter Barker, and Xiang Chen. 2006. *The Cognitive Structure of Scientific Revolutions*. Cambridge: Cambridge University Press.

Bird, Alexander. 2001. *Thomas Kuhn*. Chesham Bucks: Acumen.

Friedman, Michael. 2008. "Ernst Cassirer and Thomas Kuhn: The Neo-Kantian Tradition in History and Philosophy of Science." *Philosophical Forum* 39(2): 239–252.

Fuller, Steve. 2000. *Thomas Kuhn: A Philosophical History for Our Times*. Chicago: University of Chicago Press.

Gattei, Stefano. 2008. *Thomas Kuhn's 'Linguistic Turn' and the Legacy of Logical Empiricism*. Aldershot: Ashgate.

Hacking, Ian. 1990. "Two Kinds of 'New Historicism' for Philosophers." *New Literary History* 21(2): 343–364.

Horwich, Paul, ed. 1993. *World Changes: Thomas Kuhn and the Nature of Science*. Cambridge, MA: The MIT Press.

Hoyningen-Huene, Paul 1993. *Reconstructing Scientific Revolutions: Thomas S. Kuhn's Philosophy of Science*. Chicago: The University of Chicago Press.

Kindi, Vasso. 2010. "Novelty and Revolution in Art and Science: The Influence of Kuhn on Cavell." *Perspectives on Science* 18(3): 284–310.

Kuhn, Thomas S. 1987. *Black-Body Theory and the Quantum Discontinuity, 1894–1912*. 2nd ed. Chicago: The University of Chicago Press.

Marcum, James A. 2005. *Thomas Kuhn's Revolution: An Historical Philosophy of Science*. London: Continuum.

Nickles, Thomas, ed. 2002. *Thomas Kuhn*. Cambridge: Cambridge University Press.

Pickering, Andrew. 1984. *Constructing Quarks: A Sociological History of Particle Physics*. Chicago: The University of Chicago Press.

Preston, John. 2008. *The Structure of Scientific Revolutions: A Reader's Guide*. London: Continuum.

Richardson, Alan. 2007. "'That Sort of Everyday Image of Logical Positivism': Thomas Kuhn and the Decline of Logical Empiricist Philosophy of Science." In A. W. Richardson and T. E. Uebel (eds.), *The Cambridge Companion to Logical Empiricism*. Cambridge: Cambridge University Press, 346–369.

Rowbottom, Darrell P. 2011. "Stances and Paradigms: A Reflection." *Synthese* 178(1): 111–119.

Shapin, Steven. 1992. "Discipline and Bounding: The History and Sociology of Science as Seen through the Externalism-Internalism Debate." *History of Science* 30: 333–369.

Shapin, Steven. 2010. "Lowering the Tone in the History of Science: A Noble Calling." In *Never Pure*. Baltimore: Johns Hopkins University Press, 1–14.

Sharrock, Wes, and Rupert Read. 2002. *Kuhn: Philosopher of Scientific Revolutions*. Cambridge: Polity Press.

Wray, K. Brad. 2011a. "Kuhn and the Discovery of Paradigms." *Philosophy of the Social Sciences* 41(3): 380–397.

Wray, K. Brad. 2011b. *Kuhn's Evolutionary Social Epistemology*. Cambridge: Cambridge University Press.

Wray, K. Brad. 2012. "Assessing the Influence of Kuhn's *Structure of Scientific Revolutions*." *Metascience* 21: 1–10.

Part I
Origins and Early Reception

2 Kuhn and Logical Positivism
Gaps, Silences, and Tactics of SSR

Gürol Irzik

That Thomas Kuhn's *The Structure of Scientific Revolutions* (SSR from here on) was the most influential work responsible for the demise of logical positivism needs no documentation. Nevertheless, Kuhn devotes surprisingly little space to an explicit discussion of the logical positivists' views that are relevant for his purposes. His remarks about them are often too general and limited to a small subset of them; his treatment of logical positivism is too monolithic, ahistorical, and full of holes as there are no explicit references to the leading logical positivists or logical empiricists such as Carnap, Reichenbach, Neurath, Schlick, Feigl, and Hempel, and Frank is mentioned only once in a footnote for his work on Einstein.[1] Furthermore, recent revisionary historical accounts of logical positivism have unearthed a number of significant similarities between Kuhn's and some of the most important logical positivists', especially Carnap's, views about science. All of this raises some interesting questions that I take up in this chapter: If the main target of SSR was the logical positivist image of science, why is it that SSR contains so little about it and why was the book so successful in undermining it? What did Kuhn take logical positivism to be and which aspects of it were badly flawed according to him?

In section 1, I argue that the primary target of SSR is not logical positivism, but the textbook image of science. Logical positivism and Popperian falsificationism are Kuhn's target in SSR only secondarily. In section 2, I carefully document Kuhn's criticisms of logical positivism and argue that they are limited to the thesis of accumulation, the theory-observation dichotomy, and confirmationism. In section 3, I point to a number of absences, silences, and gaps in Kuhn's book, which weaken its attack against logical positivism. To compensate for them, Kuhn employs some tactics that I discuss in section 4. I conclude by making some suggestions for further research.

1. THE TARGET OF SSR

SSR famously opens with the sentence "History, if viewed as a repository for more than anecdote or chronology, could produce a decisive transformation in the image of science by which we are now possessed" (1). Since SSR was given the lion's share for the dethronement of the logical positivist image of science from the very beginning, that image was naturally taken to be its main target not only in later widely read texts of philosophy of science such as Suppe (1977a, 1977b) and Chalmers (1982), but also in the early reviews and discussions of SSR (Hesse 1963; Shapere 1964, 1966). While SSR did attack the logical positivist picture of science as Kuhn understood it, its target was much broader. It also included what Kuhn calls "the textbook image of science" elsewhere and, to a lesser degree, Popperian falsificationism.[2] In fact, a careful reading of SSR reveals that Kuhn's critical forays are directed more to the textbook image of science than to logical positivism (or to Popperian falsificationism for that matter). Thus, the primary target of SSR is the textbook image of science, and logical positivism is SSR's target only secondarily, or so I shall claim. Furthermore, as we shall see in section 4, Kuhn never clearly separates his secondary target from his primary target; on the contrary, he deliberately employs a strategy of associating and amalgamating, as I explain in section 4. For that reason SSR's target seems somewhat nebulous.

The most important component of the image of science Kuhn attacked in SSR, as the introductory section[3] reveals early on, was "the development-by-accumulation" view, that is, the view that science "develop[s] by the accumulation of individual discoveries and inventions" (2). Now, what is surprising is that Kuhn does not at all mention logical positivism as representing the accumulationist view in that section. Instead, he writes several times that that view has been produced by the old historiography of science and by scientists in classical science texts and science textbooks. Indeed, the very sentence that follows the opening sentence of SSR quoted in the previous paragraph makes this point: "That image has previously been drawn, even by scientists themselves, mainly from the study of finished scientific achievements as these are recorded in the classics and, more recently, in the textbooks from which each new scientific generation learns to practice its trade" (1). In short, the accumulationist view has been produced above all by "the textbook image of science" in the hands of scientists and historiographers working in the old tradition.

Kuhn discusses the place and the function of science classics, textbooks, and pedagogy throughout his book. We have already seen how central they are in the introductory section of SSR. Kuhn also talks about their role in the establishment of normal science in section II. He writes: "[Science] textbooks expound the body of accepted theory, illustrate many or all of its successful applications with exemplary observations and experiments" (10). However, they seldom recount such achievements in their original form (ibid.). They often start from first principles and justify the concepts they introduce (20). In sections VIII and IX, Kuhn remarks that science textbooks make puzzle solving look like confirmation (80) and present an accumulationist view of science (108). Finally, Kuhn devotes the entirety of section XI to the textbook image of science, according to which science develops by incremental accumulation, thereby making scientific revolutions virtually invisible. Here are some typical quotations, and certainly more can be produced:

> No wonder that textbooks and the historical tradition they imply have to be rewritten after each scientific revolution. And no wonder that, as they are written, science once again comes to seem largely cumulative. (138)

> The result is a tendency to make history of science look linear or cumulative, a tendency that even affects scientists looking back at their own research. (139)

> From the beginning of scientific enterprise, a textbook presentation implies, scientists have striven for the particular objectives that are embodied in today's paradigms. One by one, in a process often compared to the addition of bricks to a building, scientists have added another fact, concept, law, or theory to the body of information supplied in the contemporary science text. (140)

Kuhn ends section XI by emphasizing the pedagogic role of classical science texts in which existing knowledge is embodied: "More than any other single aspect of science, that pedagogic form has determined our image of the nature of science and of the role of discovery and invention in its advance" (143).

It is also worth pointing out that Kuhn's 1961 article "The Function of Measurement in Modern Physical Science," reprinted in his (1977), is also a critique of the textbook presentation of scientific measurement in particular and the textbook image of science more generally. More revealingly, Kuhn writes in a footnote that "this phenomenon is examined in more detail in my monograph, *The Structure of Scientific Revolutions*, to appear when completed as vol. 2, no. 2, in the *International Encyclopedia of Unified Science. Many other aspects of the textbook image of science*, its sources and its strengths, are also examined in that place" (Kuhn 1977, 181n3; my emphasis). For Kuhn, then, we owe our image of science more to science texts and science pedagogy than to logical positivism or Popperian falsificationism. That is why they have such a central place in his book.

I shall discuss Kuhn's treatment of logical positivism in SSR in the next section. I now turn to Popperian falsificationism as one of his main targets. After criticizing theories of probabilistic verification, Kuhn takes up Popper's falsificationism in section XII. In that section he praises Popper for drawing attention to the role of "anomalous experiences" and points to a similarity between the way falsification and anomalies function in science;

by evoking a crisis, both trigger the search for an alternative theory or paradigm. Nevertheless, according to Kuhn, anomalous experiences are not the same thing as falsifying instances, and, likewise, a theory's failure to pass a test is hardly a reason for its rejection. For him every theory is born into an ocean of anomalies:

> As has repeatedly been emphasized before, no theory ever solves all the puzzles with which it is confronted at a given time; nor are the solutions already achieved often perfect. On the contrary, it is just the incompleteness and imperfection of the existing data-theory fit that, at any time, define many of the puzzles that characterize normal science. If any and every failure to fit were ground for theory rejection, all theories ought to be rejected at all times. (146)

Kuhn's criticism is of course closely connected with his notion of normal science. Normal science is a paradigm-governed activity of puzzle-solving, which cannot be seen as testing the paradigm. For Kuhn, even when puzzles resist persistent attempts for solution and thus turn into anomalies and trigger a crisis, scientists do not treat them as falsifying instances. These are points Kuhn has already made in section VIII. Interestingly, even though it is pretty clear that his main target in that section is Popperian falsificationism, he makes no mention of Popper at all. This is very typical of SSR; it rarely mentions explicitly the works and philosophers it criticizes. Here are some examples:

> Normal science does and must continually strive to bring theory and fact into closer agreement, and that activity can easily be seen as testing or as a search for confirmation or falsification. Instead, its object is to solve a puzzle for whose very existence the validity of the paradigm must be assumed. Failure to achieve a solution discredits only the scientist and not the theory. Here, even more than above, the proverb applies: "It is a poor carpenter who blames his tools." (80)

> Though they may begin to lose faith and then to consider alternatives, they do not renounce the paradigm that has led them into crisis. They do not, that is, treat anomalies as counter-instances, though in the vocabulary of philosophy of science that is what they are. . . . Once it has achieved the status of paradigm, a scientific theory is declared invalid only if an alternate candidate is available to take its place. No process yet disclosed by the historical study of scientific development at all resembles the methodological stereotype of falsification by direct comparison with nature. . . . The decision to reject one paradigm is always simultaneously the decision to accept another, and the judgment leading to that decision involves the comparison of both paradigms with nature and with each other. (77)

I think there is one more place in which Kuhn has Popper in mind as his target. This is where Kuhn talks about ad hoc changes in theory. He points out that when confronted with a conflict between theory and experience, scientists routinely devise ad hoc modifications of their theory to save it from falsification (78). Thus, the historical record shows that Popper's method-ological dictum "avoid making ad hoc moves to save a theory from falsifica-tion" is not observed by scientists in practice. Again, Kuhn makes no explicit references to Popper, but his footnote on page 146 in section XII does refer to the first four chapters of *The Logic of Scientific Discovery*, where Pop-per criticizes ad hoc modifications. In all likelihood, then, Kuhn is directly responding to Popper's claim that "scientists do not usually proceed in this way," that is, that they don't usually evade falsification by introducing ad hoc auxiliary hypotheses (Popper 1968, 42). Kuhn is saying that they often do. Let me now turn to Kuhn's critique of logical positivism.

2. LOGICAL POSITIVISM IN SSR

What is Kuhn's conception of logical positivism and how does Kuhn relate it to the accumulationist view, which is produced and sustained above all by the textbook image of science?

The first explicit mention of logical positivism and its association with the accumulationist view occurs rather late in the book, in section IX on page 98, to be exact. There Kuhn argues that only anomalies that resist accommodation into the existing paradigm give rise to new theories. From this he concludes that the new theory must be logically incompatible with the old one since the new theory must yield at least some predictions that are different from the old paradigm or theory. As an example, he gives the relationship between Newton's and Einstein's theories. According to Kuhn, the former cannot be logically included in the latter as a special case because "Einstein's theory can be accepted only with the recognition that Newton's was wrong" (98).[4]

It is in this context that Kuhn mentions logical positivism explicitly for the first time in his book. The interpretation of scientific theory, says Kuhn, "closely associated with early logical positivism and not categori-cally rejected by its successors, would restrict the range and meaning of an accepted theory so that it could not possibly conflict with any later theory that made predictions about some of the same natural phenomena" (98). He reiterates this point two pages later and refers to Richard Braithwaite's *Scientific Explanation* for support (100n4). Kuhn's claim is that logical positivists try to evade the logical incompatibility between two compet-ing theories such as Newton's and Einstein's by restricting the domain of application of the former theory. In this way, from the logical positivists' perspective, Kuhn contends, Newton's theory is still considered valid and thus can be seen as a special case of Einstein's theory for low velocities.

He then goes on to argue against such a restriction on two grounds. First, it would immunize theories from facing any anomalies or crises, in a way reminiscent of Popper's point that such a move would render a theory immune from refutation; second, the derivation of Newton's theory from Einstein's would be spurious since the meanings of mass, time, and space in the two theories are not the same. Pursuing the latter objection to its logical conclusion, Kuhn arrives at his famous meaning incommensurability thesis (99–103).

What concerns me here is not whether these are good objections, but rather the way Kuhn associates logical positivism with the accumulationist view. To begin with, note that Kuhn attributes that view to logical positivism not directly but indirectly. In other words, he gives neither any names nor any references to any works that provide any textual evidence to the effect that logical positivists held an accumulationist view of science. Rather, he associates the latter with the former in a roundabout way by attributing to the (early) logical positivists a certain interpretation of scientific theory.[5] Kuhn implicitly suggests that the reason why (early) logical positivists interpreted scientific theories so as to restrict their domain of application was to protect the accumulationist view of science, which they supposedly believed from early on. But who are those "early logical positivists" who endorsed the interpretation of scientific theory in question and who are their successors who did "not categorically reject" it? Kuhn does not say. Furthermore, where is the evidence that shows that logical positivists indeed advocated or endorsed the accumulationist view of science? Curiously, all Kuhn can provide as evidence for the accumulationist view is a reference to Philip Wiener's review of his earlier book on the Copernican revolution (see 98n2). In that review Wiener complains that Kuhn asserts without any argument that "only the list of explicable phenomena grows; there is no cumulative process for the explanations themselves. As science progresses, its concepts are repeatedly destroyed and replaced, and today Newtonian concepts seem no exception" (Kuhn 1957, 265). Wiener then comments:

> Now, I should think that relativity theory does not destroy the Copernican and Newtonian concepts, if "destroy" means eliminating them completely along with their confirmatory evidence. Relativity theory corrects them by enlarging the frame of reference to any part of the heavens and by the metric invariance of the Lorentz transformations. In the special cases where the relativity correction is smaller than the degree of precision of our measuring instruments, scientists still make valid use of pre-Einsteinian and even pre-Copernican astronomy. In the light of this logical continuity between the limited data on which Copernicus built his carefully detailed system and the enlarged framework and more precise data of modern astronomy, the Copernican explanation is part of the cumulative growth of scientific explanations in astronomy. (Wiener 1958, 298)

Wiener was one of the founders of *The Journal of History of Ideas*, served as its executive editor from 1955 to 1986, and wrote about the history of pragmatism, the history and philosophy of science, and Leibniz. A number of his papers appeared in the journal *Philosophy of Science* as well. While he was a prolific philosopher whose long career covered a period of nearly fifty years from the 1930s well into the 80s, he was no logical positivist. Thus, the most that the reference to Wiener's paper supports is that the accumulationist view was held by some philosophers of science in the late 50s, but it does not show that it was held by the logical positivists.

In short, in the section in which Kuhn mentions logical positivism explicitly for the first time in his book, he provides no evidence for his claim that logical positivists advocated the accumulationist view. Similarly, he provides no evidence for his claim that they endorsed an interpretation of scientific theories which would restrict the domain of application of the earlier accepted theory so that it could not conflict with any later theory. Kuhn merely assumes that logical positivists held on to an accumulationist view of science and then attributes to them a certain view of scientific theories that function to protect the accumulationist view from "refutation."

To avoid any misunderstanding, let me point out that I am not claiming that no logical positivist held the accumulationist view. All I have argued up to this point is that, so far in his book (which is roughly two-thirds of the way), Kuhn has not been able to provide his readers with any reason to think that logical positivists were accumulationists.[6]

There is of course an obvious alternative way of doing this, namely, through the notorious theory-observation (T/O) dichotomy. Since, according to this dichotomy, observation is independent of any theory, observational findings and the empirical generalizations based on them do not get affected by theoretical upheavals, so they naturally accumulate through time. Already in the introductory section Kuhn signals that he would reject any categorical T/O distinction. Without mentioning any philosophers or philosophical movements, he writes: "Scientific fact and theory are not categorically separable, except perhaps within a single tradition of normal-scientific practice" (7). Later in the book, when he discusses the nature and necessity of scientific revolutions in section IX, he indeed links the T/O distinction with the accumulationist view: "In section X we shall discover how closely the view of science-as-cumulation is entangled with a dominant epistemology that takes knowledge to be a construction placed directly upon raw sense data by the mind" (96). In that section Kuhn tells us that the dominant epistemology in question is the Cartesian one and expresses doubts about categorically separating raw sensory data from theories. He takes up this issue at two levels: the level of perception and the level of language. At the perceptual level, he appeals to the famous example of the duck-rabbit in gestalt psychology. The point of the example is that two people with the same retinal images do not necessarily see the same things. At the level of language, he observes that all attempts to formulate a pure observation language have so far failed. In both cases, his criticisms are expressed rather cautiously:

> But is sensory experience fixed and neutral? Are theories simply man-made interpretations of given data? The epistemological viewpoint that has most often guided Western philosophy for three centuries dictates an immediate and unequivocal, Yes! In the absence of a developed alternative, I find it impossible to relinquish entirely that viewpoint. Yet it no longer functions effectively, and the attempts to make it do so through the introduction of a neutral language of observations now seem to me hopeless. . . . As for a pure observation-language, perhaps one will yet be devised. . . . No current attempt to achieve that end has yet come close to a generally applicable language of pure percepts. (126–127)

As before, Kuhn does not explicitly mention logical positivism as the target of his criticisms here; he assumes that it is understood. Thus, logical positivists are claimed, by allusion, to be holding a pure or neutral observation language untainted by concepts and theories. In this way logical positivists are implicated with the accumulationist view via their supposed endorsement of a sharp T/O distinction. This is as close as Kuhn gets to establishing a link between logical positivism and the accumulationist view in his entire book.

Another topic Kuhn addresses in SSR in relation to logical positivism is confirmation or, as he calls it, verification. He takes up this topic in two sections: in section VIII, "The Response to Crisis," and in section XII, "The Resolution of Revolutions."

As we saw, in section VIII Kuhn's main target is Popperian falsificationism, but in passing he also criticizes the view that sees the attempts to bring a theory and fact into agreement as confirmation. Normal science, says Kuhn, does bring the paradigm and facts into closer agreement, but this is best understood as an activity of puzzle solving rather than testing the paradigm that results either in confirmation or refutation (80).

In section XII, on the other hand, Kuhn turns to "probabilistic theories of verification." One expects that he would finally take on the arch-positivist Rudolf Carnap here and drill *The Logical Foundations of Probability*, published twelve years before SSR, but instead he refers, in a footnote, to Ernest Nagel's brief monograph *Principles of the Theory of Probability*, published in the International Encyclopedia of Unified Science series back in 1939. "Two of the most popular contemporary philosophical theories about verification," both of which are probabilistic, writes Kuhn, "all have recourse to one or another of the pure or neutral observation languages discussed in section X. One probabilistic theory asks that we compare the given scientific theory with all others that might be imagined to fit the same collection of observed data. Another demands the construction in imagination of all the tests that the given scientific theory might conceivably be asked to pass" (145). Thus, Kuhn's first dissatisfaction with probabilistic theories of verification is that they all presuppose a neutral observation language, which he has criticized earlier. His second dissatisfaction concerns a practical difficulty that such theories face. One such theory requires that given a scientific theory and a set of data, the theory in question be compared with all other scientific theories that fit the same data; another requires that given a scientific theory all possible tests of it be considered. Otherwise, probabilities cannot be computed according to either of them. Kuhn finds these requirements not only impossible to fulfill, but also misguided: "Verification is like natural selection: it picks out the most viable among the actual alternatives in a particular historical situation"(146). Thus, according to Kuhn, verification should not be seen as establishing the agreement of fact with theory. History of science shows that any major scientific theory that has gained the confidence of the relevant scientific community for some time has agreed with facts more or less. If, however, the point is put in a comparative fashion in the actual, historical circumstances, then it has merit: "It makes a great deal of sense to ask which of the two actual and competing theories fits the facts better" (147).

In contrast to the logical positivists and Popperian falsificationists, Kuhn believes that verification and falsification are not mutually exclusive processes, but rather complementary. This is because falsification is not a two-cornered fight between fact and theory but a three-cornered one between fact and two competing theories; thus, falsification is a "subsequent and separate process that might equally well be called verification since it consists in the triumph of a new paradigm over the old one. Furthermore, it is in that joint verification-falsification process that the probabilist's comparison of theories plays a central role" (147).

There is a final issue addressed in SSR that is relevant to our purposes, and it is the place and role of rules in scientific research. Kuhn devotes much of section V, "The Priority of Paradigms," to this topic and argues that the determination of shared paradigms is not the same thing as the determination of shared rules. According to Kuhn, symbolic generalizations, concepts, and particular problem solutions are never acquired in the abstract by learning rules, but rather through their applications, which enable the scientist to inspect a paradigm directly, so to speak. Once a paradigm is thus acquired, normal scientific activity can be carried out without any set of shared rules and can proceed "without agreeing on or even attempting to produce a full *interpretation* or *rationalization*" of a paradigm (44; emphasis in original). By studying model solutions to typical problems (i.e., paradigm in the sense of exemplar as Kuhn dubbed it later in "The Postscript"), scientists acquire an ability to see resemblances between solved problems and unsolved ones and thus apply the paradigm to the latter without the help of any rules. For this reason, much of scientists' knowledge is "tacit," that is, inarticulable in terms of rules or propositions, a point Kuhn owes to Ludwig Wittgenstein and Michael Polanyi (44–46).

Although Kuhn does not mention logical positivism explicitly in this context, it appears that that is in part his target.[7] For logical positivists, rules play a role in science at least in two ways. First, there are rules of confirmation which determine the relationship between theory and evidence, or, more specifically, which specify the degree to which a given body of evidence supports a theory. Second, as is well known, a number of logical positivists, including Carnap, Hempel, and Feigl, defended a conception of scientific theory in terms of the conjunction of theoretical postulates and correspondence rules, a conception dubbed as the "orthodox" or the "received" view (see, for example, Carnap 1956 and Feigl 1956). Leaving aside the logical and mathematical terms, the language of a scientific theory is bifurcated into two classes, an observation language and a theoretical language, consisting of observational terms and theoretical terms respectively. Theoretical postulates, which express the fundamental principles of the theory, contain only theoretical terms, which acquire their empirical meaning or content through the correspondence rules. The latter connect theoretical terms to observational ones and thereby partially interpret the former and facilitate the application of the theory to particular phenomena.

As we saw above, Kuhn expressed his skepticism about the usefulness of the confirmation rules in practice in section XII even if they were clearly formulatable. Thus, it is the rules of correspondence that Kuhn seems to be attacking here. He is arguing that scientists do not apply a theory to phenomena through such rules and that a theory in the sense of a paradigm and its applications form a whole that cannot be separated from each other. In quite Wittgensteinian spirit, he is saying that the meaning of terms such as "mass," "force," and "space" are learned much less through any explicit definitions or rules than through their applications to particular problem solutions (see SSR, 46–47).

This completes my documentation of the place of logical positivism in SSR. Let me conclude this section by noting that there are a hundred and fifty footnotes to SSR, and only about a dozen of these refer to philosophers, and astonishingly, only two of those philosophers can be associated with logical positivism: Braithwaite and Nagel.[8] Of the two, only Nagel's work is criticized, but even then the criticisms are limited to his brief monograph on probabilistic theories of confirmation, as we just saw. Other logical positivists are never mentioned explicitly and indeed ignored almost completely. It is to these absences that I now turn.

3. ABSENCES AND SILENCES IN SSR: THE KUHN PUZZLE

In his book *A Theory of Literary Production*, the French literary critic Pierre Macharey, a former student of Louis Althusser, argued that what a text does *not* say is as important as what it does say. He wrote: "It is in the significant silences of a text, in its gaps and absences, that the presence of ideology can be most positively felt" (Macharey 1978, 60). Thus, we can speak of an "ideology of a text" that, in a quite Wittgensteinian sense, shows itself in its gaps, absences, and silences, which typically operate behind the back of the author, so to speak, and therefore need not be conscious at all. Macharey developed his theory of ideology for interpreting literary texts, but it might also be applied to at least some philosophical texts, certainly in my view to Kuhn's SSR.

George Reisch's groundbreaking book, with the revealing title *How the Cold War Transformed Philosophy of Science*, documented how logical positivism was reduced to an apolitical and technical movement concerned only with "the logic of scientific inquiry," stripped of all of its progressive social and political aspects, under the ideology of the Cold War in the United States in the 50s. SSR is both a product of and a contributor to this ideological transformation (Reisch 2005; see also Fuller 2000). As we saw, it presents a very narrow picture of logical positivism, not only confining itself to the thesis of accumulation, the T/O dichotomy, and confirmationism, but also eschewing the protocols debate and the antifoundationalism that results from it. It says nothing about the social and political agenda of the logical positivist movement, it is blind to different trends within it, and despite popular perception to the contrary, it is written from quite an internalist perspective that contains little about the social factors that influence the development of science. SSR can be said to reflect an ideological orientation through its gaps, absences, and silences and thereby to legitimate almost a caricature of logical positivism. Below I give some of the examples of these that are most relevant for the purposes of this chapter.

The most conspicuous absence in SSR is any reference to Rudolf Carnap, the most influential of the logical positivists both in Europe before the war and in the United States after the war. Notwithstanding the discussion about the theory-observation distinction, SSR contains absolutely no mention of any of Carnap's views. Even in the context of that discussion, Carnap's name does not appear anywhere. Moreover, somewhat ironically, Carnap held several views that are similar to Kuhn's, as the recent revisionary work on the history of logical positivism has revealed. This is not the place to document them, but suffice it to say that both of them endorsed a moderate semantic holism and a form of semantic incommensurability, and both rejected foundationalism in epistemology; there are also striking similarities between their characterizations of scientific revolutions as well as between the functions Carnap attributed to scientific theories and the functions Kuhn attributed to paradigms. These similarities leap to the eye especially after Kuhn begins to formulate his views in terms of taxonomic lexicons.[9]

Toward the end of his life, Kuhn publicly confessed that he had not read any of Carnap's post-Aufbau writings and admitted acting irresponsibly (Kuhn 1993, 313; Kuhn 2000, 305–306). This explains why Carnap is missing from SSR, though it does not justify it. After all, SSR did attack logical positivism as Kuhn understood it, and Carnap must surely have figured in any understanding of logical positivism.

It is not only Carnap that is missing from SSR, but also Hans Reichenbach and Otto Neurath, whose boat surely must have sailed in Kuhn's intellectual waters through Quine. SSR contains nothing about the rejection of antifoundationalist epistemology signified by Neurath's boat metaphor. Hans Reichenbach, on the other hand, makes only a brief appearance in SSR, if not in name certainly through his famous context of discovery-context of justification distinction, which Kuhn says his book might have violated (8). However, it is to be noted that even though he did read *Experience and Prediction* (Kuhn 2000, 283), he is oblivious to the fact that Reichenbach emphasized the "gestalt character" of our experiences in his book: "We do not see things as amorphous but always as framed within a certain description. . . . In the same sense the objects of our sensations have always a 'Gestalt character.' They appear as if pressed into a certain conceptual frame; it is their being seen within this frame which we call existence" (Reichenbach 1938, 221). Thus, as early as 1938 Reichenbach denied "the myth of the given" and pointed out the concept-ladenness of experience, much as Kuhn did years later in SSR. This is not to say that there are no differences between their uses of the gestalt character of our experiences; for example, Kuhn's view that paradigm changes are like gestalt switches, which enable the perceiver to see "the same thing" as two different things, is absent in Reichenbach's philosophy. Nevertheless, the fact that Reichenbach was not a crude positivist who succumbed to the myth of the given is a point Kuhn missed. All of these similarities diminish the epistemological distance between Kuhn and logical positivism so far as perception is concerned; or, conversely, their absence makes that distance bigger than it actually is.

Another important absence in SSR is Philip Frank. As is well known, Frank was one of the major representatives of logical positivism after the war as well as before, therefore both in Europe and in the United States, where logical positivism flourished most from the end of the war to the early 60s. Herbert Feigl, an influential figure in the logical positivist movement in the United States, delineated three major trends within the logical positivist movement and cited Frank as representing one of them:

> As the movement of logical empiricism attained its world-wide scope, we may clearly discern three major trends differing amongst each other more in their method of procedure than in their basic outlook. There is first the trend exemplified most typically by the work of Philipp Frank (and to some extent also by the earlier work of Neurath and von Mises), which combines informal logical analyses of the sciences with a vivid awareness of psychological and social-cultural factors operating in the selection of problems and in the acceptance or rejection of hypotheses and which contribute to the shaping of certain styles of

scientific theorizing. In a sense, this is a genuine sequel to the work of Ernst Mach. (Feigl 1956, 4–5)

Frank held a number of views that are similar to Kuhn's (see Frank 1949, 1956, 1957). Those most relevant for our purposes are as follows. Frank believed that a logico-structural analysis of science was not sufficient for a deep understanding of science and must be supplemented by a pragmatic-historical one. He thus emphasized the importance of history of science for doing philosophy of science and often appealed to historical cases such as the Copernican revolution for insight into scientific development. He paid particular attention to the issue of theory acceptance and noted, much as Kuhn did, that theory acceptance is not just a matter of match between theory and observational facts since no theory ever agrees with all facts completely, but only more or less. In addition to agreement with facts, he also underlined the importance of the criterion of simplicity (understood both as mathematical simplicity and as simplicity of the overall theoretical discourse) and drew attention to the problem of how to weigh these two factors in choosing among rival theories, an issue that Kuhn dealt with especially in his influential 1973 article "Objectivity, Value Judgment and Theory Choice," reprinted in *The Essential Tension*.[10]

Now, despite the fact that Kuhn was at Harvard when Frank taught there and despite the fact that he attended his lectures and read him (Kuhn 2000, 305), SSR contains no mention of his views, which I have cursorily summarized above. In SSR Kuhn refers to Frank only once, and that is to his book on Einstein in the context of lay reactions to the notion of curved space in the theory of relativity (see 149n3). Nor is there any reference, much less any expression of indebtedness, to Frank's works in Kuhn's later writings, not even in his 1973 article, in which the similarity becomes particularly conspicuous.[11]

How is Kuhn's total neglect of Frank's ideas to be explained, leaving aside simple forgetting? Several answers, admittedly sketchy and speculative, can be offered. Perhaps Kuhn was not moved by those passages in Frank's books, since they are embedded in a philosophical perspective that is different from his. After all, Frank was a representative of logical positivism, which Kuhn aimed to dismantle. It also sometimes happens that when we come across ideas similar to ours, we make them our own without ever realizing it. Perhaps that is what happened to Kuhn. Conversely, we tend to exaggerate our differences from others whose ideas we are reacting against in order to emphasize the originality of our own. Any one of these might have been the case with Kuhn.

There are, however, more puzzling kinds of absences in Kuhn's book that are difficult to explain. I will mention two such examples, both of which have the same function: their neglect weakens the case Kuhn wants to make against the logical positivist thesis of accumulation. Recall that in section 2 I argued that Kuhn had difficulty linking the thesis of accumulation with

logical positivism and that the only argument he provided was through the
T/O dichotomy. If observation is independent of any theory, then changes
at the latter level do not affect the former, so observation reports and low-
level empirical laws that refer to observables can naturally accumulate,
and in this way science becomes an incrementally accumulative enterprise.
A number of logical positivists indeed believed something like this. For
example, in a frequently quoted paper, Feigl pictured the empirical (that is,
nontheoretical) concepts as closely tied to the "soil" of observation/experi-
ence and argued that "indeed, in the view of Carnap, the O. L. [observation
language] is not in any way theory-laden or 'contaminated' with theoretical
assumptions or presuppositions" (Feigl 1956, 7).

However, the T/O dichotomy gives rise to a rather weak form of accumulationism since by itself it does not make possible any accumulation
at the theoretical level. There is a stronger form of the accumulationist
thesis which allows for just that, and there are two ways of linking the logical positivists with the stronger version: through Ernest Nagel's thesis of
theory reduction and Carl Hempel's deductive-nomological (D-N for short)
model of explanation (see Nagel 1949, 1961; Hempel 1948).

According to a common version of the thesis of theory reduction, a theory
A is reducible to another theory B if the laws of A are derivable from B under
certain conditions. When A is reduced to B, A becomes a special case of B and
cannot possibly conflict with B. Therefore, if A is a highly confirmed theory
and thus accepted by the scientists, it stays that way with the invention of B.
Of course, the excess content of B must in turn be tested, but assuming it is
also confirmed and thus accepted, science displays an accumulative progress
from A to B at the theoretical level. Indeed, theory reduction is seen as a major
process by which science grows in a cumulative manner for the advocates of
the so-called "received view of scientific theories" (see, for example, Suppe
1977a, 53–56). Furthermore, A and B must share a number of descriptive
terms and statements that have the same meanings in some types of reduction
such as those that are called "homogenous" (see Nagel 1961, 338–358). Thus,
theory reduction requires meaning invariance at least in those cases.

Turning to the D-N model of explanation, an event (the explanandum)
is explained by deducing it from a law or a set of laws with appropriate
particular facts (explanans), and a law is in turn explained by deducing it
from an even more general law or laws. Now, if this is indeed how laws are
explained in science, then the laws in the explanandum can never conflict
with the laws in the explanans. In other words, genuine scientific explanations always preserve previously accepted laws. We have then accumulation
at the level of theories, not just at the observational level. As Frederick Suppe
has observed, this is a form of intertheoretic reduction (Suppe 1977b, 621).
Thus, the thesis of theory reduction and the D-N model of explanation are
interrelated, and both give support to the accumulationist view.

As is well known, the D-N model of explanation is synonymous with
Hempel's name, and Hempel, a shining student of Reichenbach, was one

of the most important representatives of logical positivism in the United States from the 50s onward. Ernest Nagel, who devoted more attention to the topic of theory reduction than anyone else, was also an influential figure who held a number of views in the spirit of logical positivism (see Nagel 1949, 1961). Here are then two routes, through Nagel's and Hempel's works, of establishing a direct link between logical positivism and the strong version of the accumulationist view, so one would expect Kuhn to make this obvious connection and then attack them. Curiously, however, this is not something Kuhn does, even though he was familiar with both Nagel's and Hempel's views (see SSR, xii, 145n1; 1977, 15–16). The D-N model of explanation is totally absent in SSR, and the puzzle regarding reduction is exacerbated by the fact that Feyerabend discussed Nagel's views about theory reduction and the requirement of meaning invariance in detail in his famous 1962 paper, which Kuhn knew well (see Feyerabend 1962, 47n6). Even though Kuhn does attack the view that Newton's theory can be seen as a special case of Einstein's, he does not bring up the thesis of theory reduction at all.[12]

I call these last two absences *the Kuhn puzzle* for obvious reasons. It is hard to explain why neither the thesis of theory reduction nor the D-N model of explanation appears anywhere in SSR. For one thing, they would have enabled Kuhn to link the accumulationist view with logical positivism in ways other than that which is provided by the T/O dichotomy. For another, their absence deprives Kuhn of undermining a stronger version of the accumulationist view with the resources already at hand in SSR. These conspicuous gaps and absences seriously weaken the case Kuhn wants to make against logical positivism.

4. THE TACTICS OF SSR

Despite a number of serious argumentative and other weaknesses in its attack against logical positivism, SSR became the most influential work responsible for logical positivism's demise. I believe the success story of SSR has yet to be told in full, and I will say more about it in the last section. Here I would like to draw attention to an aspect of Kuhn's book that has not been sufficiently emphasized, if at all: it is the tactics of association and amalgamation Kuhn skillfully employs in his book, tactics that compensate for its weaknesses and magnify its force greatly.

As we saw in section 1, Kuhn aims to destroy the textbook and the logical positivist images of science all at once. As a physicist who has turned into a historian of science for philosophical purposes, he is well trained in science texts and history of science, but has very limited knowledge of logical positivism. In order to compensate for this lack of knowledge and to boost the effect of his critique, he at times associates and at other times amalgamates the image of science portrayed by the old historiography of science and by scientists in classical science texts and science textbooks with certain aspects of logical positivism. Let me give three examples.

First, recall how SSR begins: by telling the reader on the very first page how the new historiography can transform the accumulationist picture of science commonly held, a picture that is the product of the textbook image of science. Two pages later, he writes that the first result that will emerge from this new historiography is "the insufficiency of methodological directives, by themselves, to dictate a unique substantive conclusion to many sorts of scientific questions" (3). Nowhere in his book does he tell us what sort of "methodological directives" he has in mind, but the phrase brings to mind rules of confirmation (like "accept a theory if it is highly confirmed") that logical positivists, especially Carnap, have supposedly developed for theory choice.[13] Why else would he repeatedly say that logic and experimentation alone cannot determine theory choice? In this way, SSR establishes an associative link in the minds of his readers between its explicit target, namely the textbook image of science, and logical positivism without ever mentioning the latter. Thus, from the very beginning, the target includes not just the textbook image of science, but also logical positivism by association.

Second, toward the end of the "Introduction," on page 8, summarizing the content of each section of his book, Kuhn tells his readers that while section XI will discuss how *the textbook tradition* makes scientific revolutions invisible, section XII will describe "the process that should somehow, in a theory of scientific inquiry, replace *the confirmation or falsification procedures made familiar by our usual image of science*" (8; my emphasis). This establishes not only another link between "our usual image of science" and logical positivism (via the phrase "the confirmation procedures"), but also a link between the textbook image and logical positivism because Kuhn had earlier written that "our usual image of science" has been produced above all by the textbook image of science. All of this contributes to the blurring of the differences between the textbook and the logical positivist images of science.

Kuhn's most explicit employment of his tactics of association and amalgamation appears in section XI. As we just saw, Kuhn's aim in that section is to show and explain how revolutions appear invisible to all except to the good historians of science. The main thesis of that section is that most people's knowledge of science is derived from science textbooks and that because of their pedagogic function, science textbooks contain nothing about the actual scientific practice and experience and are written from the perspective of normal science. For that reason, Kuhn contends, they portray an accumulative picture of scientific development, making revolutions virtually invisible. Now, although the entire section is devoted to displaying this function of science textbooks, Kuhn holds responsible for this distorted view of scientific development not only them, but also "the popularizations and the philosophical works modeled on them":

As the source of authority, I have in mind principally textbooks of science together with the popularizations and the philosophical works modeled on them. All three of these categories . . . have one thing in common. They address themselves to an already articulated body of problems, data, and theory, most often to the particular set of paradigms to which the scientific community is committed at the time they are written. Textbooks themselves aim to communicate the vocabulary and syntax of a contemporary scientific language. Popularizations attempt to describe these same applications in a language closer to that of everyday life. And *philosophy of science, particularly of the English-speaking world, analyzes the logical structure of the same completed body of scientific knowledge* [my emphasis]. . . . All three record the stable *outcome* of past revolutions and thus display the bases of the current normal-scientific tradition. (136–137)

No logical positivists are explicitly mentioned here, but enough is said ("philosophy of science in the English-speaking world, analyzing the logical structure of knowledge") to bring them to mind. This is the only place where Kuhn mentions this kind of philosophy of science in that section, which is obviously an allusion to logical positivism. The rest of the section smoothly returns to the theme of the nature and role of textbooks in science and discusses it exclusively. In this way, the textbook image of science, its popularizations, and logical positivism are all lumped together, all allegedly having the same function of hiding scientific revolutions and presenting an image of science that is accumulationist. Kuhn's tactics of association and amalgamation could not have been more apparent.[14]

5. CONCLUDING REMARKS

Logical positivism in the 50s in the United States was quite different from the logical positivism in Europe before the war. Kuhn was reacting against a version of the former that he called "that sort of everyday image of logical positivism," which made all of the neo-Kantian roots and concerns of prewar logical positivism invisible (Richardson 2007). There is little work about what logical positivism was and how it was perceived by philosophers in the United States from the 40s to the early 60s.[15] Although this is a topic that requires a study of its own, I would like to make one general point that may help us with the framing of such an inquiry. It is that we should cast our historical net as wide as possible. I do not just mean that the inquiry must also include less influential or largely forgotten figures, such Gustav Bergmann and Richard Braithwaite and others, who have contributed to the journal *Philosophy of Science*. Nor do I simply mean that it should look into not only the primary, but also the secondary literature, such as book reviews and general surveys of philosophy of science written in that period. Such research should certainly be carried out. I can perhaps express what I have in mind better by drawing attention to Herbert Feigl's overview of the major trends of logical positivism in his 1956 article, from which I have already quoted in section 3. Recall that in that quotation, Feigl first mentions Frank's approach to science, which combines informal logical analysis with a sociohistorical one, as one major trend. The rest of that quotation goes as follows:

> There is, secondly, the trend characterized variously as "analytic phi-losophy," "therapeutic positivism," or "casuistic logical analysis," originally introduced by G. E. Moore in England and most strikingly developed and modified by Wittgenstein. Nowadays Cambridge and Oxford, the second even more strongly than the first, are among the chief centers of this type of philosophizing. Here we find the Socratic method applied with extreme subtlety to the peculiarities (ambigui-ties and vaguenesses, strata and open horizons, implicit rules) of natu-ral languages. To be sure, extreme preoccupation with this approach has led to some excesses which were stigmatized as "futilitarian" (the phrase is Gustav Bergmann's). Nevertheless, this informal but often very brilliant method is fundamentally not as different as it may appear from either the older (first mentioned) positivistic-pragmatist approach, or the more rigorous logical reconstruction method. This last and third method is best exemplified in the work of Carnap and Reichenbach, and in that of their students. It is also pursued in the work of Woodger and Braithwaite, as well as in the work of Tarski, Mehlberg and others among the Polish logicians and methodologists. (Feigl 1956, 4–5)

I think it is truly extraordinary that Feigl mentions "analytic philosophy" as represented by Moore and later Wittgenstein as a major trend in logical positivism alongside Frank's, Carnap's, and Reichenbach's approaches. His surprising classification may serve as a valuable warning for any tendency to cast the historical net too narrowly.[16]

I would like to further suggest that the net should include not just the representatives of and the sympathizers with logical positivism, but also its critics. This is because in criticizing the views of her opponents a critic often contributes to the shaping of the readers' understanding of them. In other words, there is always a dialectical interplay between interpretive reconstruction, which is not always faithful to the original, and criticism. Consider, for example, Quine's critique of Carnap in his famous "Two Dogmas of Empiricism" paper (1951). His critique of the *Aufbau* project in that paper has at the same time served as an influential history of a significant chapter of logical positivism, a history from which most of us "learned" that Carnap's project was a foundationalist one virtually indis-tinguishable from Russell's and that it aimed to reduce all scientific knowl-edge to incorrigible propositions of sense-data (see Creath 2007). The fact that Quine got Carnap's project entirely wrong was beside the point; many read Quine's critique of the *Aufbau* rather than the *Aufbau* itself, which became available in English only in 1967.

As another example, take Feyerabend's critique of "the positivistic interpretation" of science, which he developed in a series of papers (Feyerabend 1958, 1960, 1962). In those papers Feyerabend distinguished between two versions of this interpretation. One of them is instrumentalism, according to which scientific theories are merely instruments for prediction and lack any descriptive meaning. According to the other version, scientific theories do possess descriptive meaning, but their meaning derives solely from their connection with observation via correspondence rules. Feyerabend attributed the second version to Carnap. He claimed that any philosopher who endorses the positivistic interpretation of science in one version or another is thereby committed to the "stability thesis," according to which interpretations or meanings of scientific terms are theory-independent. He then went on to criticize these views, develop a realist alternative, and arrive at his famous thesis of meaning variance.

Quine's and Feyerabend's papers were influential not only as critiques of logical positivism, but also in shaping what came to be "the received view" of it. Although we know today that both Quine's and Feyerabend's characterizations of Carnap's views in these papers are seriously flawed (see Friedman 1999 and Richardson 1998), this does not change the fact that, in all likelihood, Kuhn's views about logical positivism were influenced by them. Kuhn and Quine were members of the Harvard Society of Fellows during the same period, and Kuhn was very much influenced by Quine's "Two Dogmas of Empiricism" (Kuhn 2000, 279). Similarly, Kuhn and Feyerabend were colleagues at Berkeley in the early 60s, and as we noted in section 3, Kuhn was also quite familiar with Feyerabend's critical papers on logical positivism, especially the 1962 paper. Given such influences, then, SSR is as much responsible for the popular understanding of logical positivism as Quine's and Feyerabend's works. As Alan Richardson perceptively remarked, "Kuhn's work helps to make widespread and to stabilize the very image of logical empiricism it sought to argue against" (Richardson 2007, 365).

Despite Kuhn's limited knowledge of logical positivism and despite a number of gaps and absences that weaken SSR's attack against it, the book was extremely effective in destroying logical positivism as it was understood then. We might even say that, paradoxically, the weaknesses of SSR contributed to its strength. For, as Kuhn said in an interview, had he known more about Carnap and logical positivism, he would have written an entirely different book (Kuhn 1994). That book would not have the holes SSR did, but then, in all probability, it would also not create the same effect.

Even though logical positivism was already in decline under the pressure of later Wittgenstein, Quine, Hanson, Feyerabend, and Putnam, the destructive impact of SSR turned out to be stronger than any of these critiques, perhaps even all of them combined. SSR, however, offered more than a critique. In retrospect, the success of SSR appears to be due less to its attack against logical positivism than to outlining an alternative "epistemological paradigm," to use Mary Hesse's phrase (Hesse 1963, 287). Skillfully and provocatively, it brought to the table a new agenda, a new language, and a new style of philosophizing. The new agenda consisted of explaining the historical development of science in which scientific revolutions figure prominently, accompanied by a set of related problems such as the dynamics and rationality of theory change, incommensurability, the nature of scientific communities, and the relationship between history and philosophy of science. SSR thus put the coffin's nail in the intractable problems of inductive logic, confirmation, and cognitive significance that preoccupied logical positivists for decades by making those problems seem irrelevant to a historical understanding of science. It also displaced the logical analysis of scientific knowledge with a novel, sociohistorical style of philosophizing, couched in a colorful and captivating language of paradigms, normal science, puzzle solving, anomalies, crises, gestalt switches, conversions, and persuasions. It inspired a whole new school of philosophy known as the sociology of scientific knowledge and even a whole new discipline called "social studies of science." We may then loosely speak of SSR initiating a kind of revolution at the philosophical level. There is evidence that Kuhn himself saw his book in this way:

> The decision to reject one paradigm is always simultaneously the decision to accept another, and the judgment leading to that decision involves the comparison of both paradigms with nature and with each other. There is, in addition, a second reason for doubting that scientists reject paradigms because confronted with anomalies or counterinstances. In developing it my argument will itself foreshadow another of this essay's main theses. The reasons for doubt sketched above were purely factual; they were, that is, themselves counterinstances to a prevalent epistemological theory. As such, if my present point is correct, they can at best help to create a crisis or, more accurately, to reinforce one that is already very much in existence. By themselves they cannot and will not falsify that philosophical theory, for its defenders will do what we have already seen scientists doing when confronted by anomaly. They will devise numerous articulations and *ad hoc* modifications of their theory in order to eliminate any apparent conflict. Many of the relevant modifications and qualifications are, in fact, already in the literature. If, therefore, these epistemological counterinstances are to constitute more than a minor irritant, that will be because they help to permit *the emergence of a new and different analysis of science* within which they are no longer a source of trouble. Furthermore, if a typical pattern, which we shall later observe in scientific revolutions, is applicable here, these anomalies will then no longer seem to be simply facts. *From within a new theory of scientific knowledge*, they may instead seem

very much like tautologies, statements of situations that could not conceivably have been otherwise. (SSR, 77–78; my emphasis)

In this extraordinary passage Kuhn draws a striking parallel between scientific paradigms and prevalent epistemological theories (shall we say "epistemological paradigms"?). Just as scientists do not reject their paradigms simply because the latter face counterinstances, neither will philosophers abandon their epistemological paradigms because of isolated counterexamples or critiques. If scientists reject a paradigm only in the presence of an alternative that is better, so will philosophers give up an epistemological paradigm if a better alternative is available. Therefore, Kuhn sees his main task in SSR not (merely) as providing more and more counterexamples to such epistemological paradigms as logical positivism and falsificationism, but as producing a paradigm that is superior to them. Moreover, he believes that he has produced a paradigm far superior to its alternatives on the grounds that it fits the historical record much better, so much so that all the facts that appear as historical anomalies for the prevalent epistemological paradigms will now seem very much like tautologies. That is no modest philosophical claim by any standard. In retrospect, we can see that Kuhn had set his hopes too high since few philosophers became strict followers of the Kuhnian epistemological paradigm. Nevertheless, SSR did change the way we philosophize about science forever. Herein lies its real greatness.[17]

NOTES

1. Kuhn does not distinguish between logical positivism and logical empiricism in his book. Neither will I since the difference between them does not matter for my purposes in this chapter.
2. Kuhn uses the phrase "the textbook image of science" in Kuhn (1977, 181n3). It appears in a footnote to his 1961 article "The Function of Measurement in Modern Physical Science," which is reprinted in his (1977).
3. SSR is divided into sections rather than chapters. Accordingly, I will follow this usage.
4. Long before Kuhn, Popper actually made the same point with respect to the relationship between Newton's theory on the one hand and Kepler's and Galileo's laws on the other in a lecture published in 1949, reprinted as an appendix with the title "The Bucket and the Searchlight: Two Theories of Knowledge" in his book *Objective Knowledge* (1975, 341–361). A revised version of the appendix became Chapter 5 of that book. See also the bibliographical note at the end of the chapter in question (Popper 1975, 204–205).
5. Commenting on Kuhn's argument, John Preston (2008, 54) claims that the interpretation of scientific theories in question is associated with an instrumentalist view of theories. This is doubtful for two reasons. First, if one is an instrumentalist, then theories, being merely tools for prediction and the organization of data, are neither true nor false, in which case the question of compatibility or incompatibility between two competing theories such as Newton's and Einstein's does not arise. So, from an instrumentalist

perspective, there is no need to restrict the domain of application of the earlier theory. Second, none of the "early" logical positivists, except Schlick at one point, was an instrumentalist.

6. Some philosophers, however, did ask whether the accumulationist view can be properly attributed to logical positivism at all (see Richardson 2007, 253ff.). Richardson rightly drew attention to the recent revisionary work that puts in doubt the view that all logical positivists were accumulationists. While it is certainly true that some of them (most notably, Carnap) rejected the accumulationist view at the theoretical level, others such as Ernest Nagel were committed to it via their endorsement of the thesis of theory reduction. Furthermore, rejecting accumulationism at the theoretical level leaves it untouched at the observational level. For more on this topic, see below and section 3.

7. I say "in part" because he may have in mind Popper as well.

8. Though not a logical positivist, Popper occupies more space than any of them. Yet, SSR contains nothing about Popper's view of scientific progress. Years later, Kuhn had this to say about it: "I was introduced to Popper at a fairly early stage, and we saw a little bit of each other. Popper was constantly talking about how the later theories *embrace* the earlier theories, and I thought that was not just going to work out quite that way. It was too positivist for me" (Kuhn 2000, 286). Despite Popper's protests, it was pretty common among philosophers back then to lump Popper together with logical positivists. It is quite likely that Kuhn too regarded Popper's philosophy of science as continuous with logical positivism.

9. I have documented these similarities in detail in (Irzik and Grünberg 1995). See also Earman (1993), Friedman (2001), and Reisch (1991). In fairness, I should point out that if Kuhn had not written SSR, it is doubtful that we would have been able to pinpoint the way Carnap characterized scientific revolutions as well as his endorsement of the theses of semantic holism and incommensurability. After all, as Kuhn said, "reading Carnap after Kuhn is not the same experience as reading him before" (personal correspondence, 6 October 1995). This correspondence is also interesting in that it contains a sincere statement of how Kuhn received the revisionary work that drew attention to the similarities between his and Carnap's views. We produce it in full as an appendix to this chapter.

10. For a fuller discussion of the similarities and differences between Frank's and Kuhn's views, see Nemeth (2007) and Reisch (2005, 229–233).

11. Adolf Grünbaum reminisced in a lecture that he and Kuhn attended a talk by Frank, who characterized scientific progress as noncumulative (Grünbaum 1996). Apparently, Kuhn was very surprised when he was reminded of Frank's talk. This story, together with Grünbaum's psychoanalytic explanation of Kuhn's total memory loss of this event, is also told in Nemeth (2007, 297–298n11).

12. I should point out that while Nagel discusses theory reduction at length, he does not give the relationship between Newton's and Einstein's theories as an example of theory reduction in his major (1961) book. His paradigm example is the reduction of thermodynamics to statistical mechanics (see Nagel 1961, Chapter 11). Kuhn could have introduced at least this example to attack the thesis of theory reduction and thereby the accumulationist view of science.

13. That the point of inductive logic as developed by Carnap was to dictate which hypothesis to choose given a body of evidence was a very common misunderstanding that pervaded the writings of many leading philosophers of science from Kuhn to Feyerabend, Lakatos and Putnam. For a critique of

this misunderstanding in the context of Carnap's views about the rationality of science vis-à-vis Kuhn's and Popper's, see Irzik (2003).

14. In this context, it may be worth remembering the following quotation by Shapere in one of the earliest reviews of SSR: "In attacking the 'concept of development-by-accumulation', Kuhn presents numerous penetrating criticisms not only of histories of science written from that point of view, but also of certain philosophical doctrines (mainly Baconian and positivistic philosophies of science, particularly verification, falsification, and probabilistic views of the acceptance or rejection of scientific theories) which *he convincingly argues are associated with that view of history*." (Shapere 1964, 383; my emphasis). I argued that there is actually no such convincing argument, but merely a skilful association. I think this is an indication of how successful Kuhn's tactics of association and amalgamation were, given that Shapere's review overall was very critical of Kuhn's own views.

15. Notable exceptions are Hardcastle and Richardson (2003) and Reisch (2005).

16. Most of us today would not think of Moore and later Wittgenstein as representing a version of logical positivism. As is well known, logical positivists were very much influenced by Wittgenstein's *Tractatus* in the 20s and early 30s, but they parted ways with him later. Feigl's classification is especially curious since Wittgenstein's later views, especially those about concepts, have been taken to undermine logical positivist theories of meaning.

17. I thank the editors of this volume for their helpful suggestions and comments.

REFERENCES

Carnap, Rudolf. 1956. "The Methodological Character of Theoretical Concepts". In H. Feigl and M. Scriven (eds.), *Minnesota Studies in the Philosophy of Science*, vol. 1. Minneapolis: University of Minnesota Press, 38–76.

Chalmers, A. F. 1982. *What Is This Thing Called Science?* Queensland: Queensland University Press.

Creath, Richard. 2007. "Vienna, the City of Quine's Dreams." In A. W. Richardson and T. E. Uebel (eds.), *The Cambridge Companion to Logical Empiricism*. Cambridge: Cambridge University Press, 332–345.

Earman, John. 1993. "Carnap, Kuhn, and the Philosophy of Scientific Methodology." In P. Horwich (ed.), *World Changes: Thomas Kuhn and the Nature of Science*. Cambridge-Massachusetts and London: The MIT Press, 9–36.

Feigl, Herbert. 1956. "Some Major Issues and Developments in the Philosophy of Science of Logical Empiricism." In H. Feigl and M. Scriven (eds.), *Minnesota Studies in the Philosophy of Science*, vol. 1. Minneapolis: University of Minnesota Press, 3–37.

Feyerabend, Paul K. 1958. "An Attempt at a Realistic Interpretation of Experience." Reprinted in *Realism, Rationalism and Scientific Method*, Philosophical Papers, vol. 1. 1981. Cambridge: Cambridge University Press, 17–36.

Feyerabend, Paul K. 1960. "On the Interpretation of Scientific Theories." Reprinted in *Realism, Rationalism and Scientific Method*, Philosophical Papers, vol. 1. 1981. Cambridge: Cambridge University Press, 37–43.

Feyerabend, Paul K. 1962. "Explanation, Reduction and Empiricism." Reprinted in *Realism, Rationalism and Scientific Method*, Philosophical Papers, vol. 1. 1981. Cambridge: Cambridge University Press, 44–96.

Frank, Philipp. 1949. *Modern Science and Its Philosophy*. Cambridge, MA: Harvard University Press.

Frank, Philipp. 1956. "The Variety of Reasons for the Acceptance of Scientific Theories." In Philipp Frank (ed.), *The Validation of Scientific Theories*. Boston: Beacon Press, 3–18.

Frank, Philipp. 1957. *Philosophy of Science*. Englewood Cliffs, NJ: Prentice-Hall Inc.

Friedman, Michael. 1999. *Reconsidering Logical Positivism*. Cambridge: Cambridge University Press.

Friedman, Michael. 2001. *Dynamics of Reason*. Stanford, CA: CSLI.

Fuller, Steven. 2000. *Thomas Kuhn: A Philosophical Fable for Our Times*. Chicago: University of Chicago Press.

Grünbaum, Adolf. 1996. "C. G. Hempel as a Philosopher of Science." Lecture delivered for the German-American Interaction in Scientific Philosophy after 1933 conference, Pittsburgh University, March 29–April 1, 1996.

Hardcastle, G., and A. Richardson, eds. 2003. *Logical Empiricism in North America*. Minneapolis: University of Minnesota Press.

Hempel, Carl G. 1948. "Studies in the Logic of Explanation." Reprinted in *Aspects of Scientific Explanation*. 1965. New York: The Free Press, 245–290.

Hesse, Mary B. 1963. Review of *The Structure of Scientific Revolutions*, by T. S. Kuhn. *Isis* 54: 286–287.

Irzik, Gürol. 2003. "Changing Conceptions of Rationality: From Logical Empiricism to Postpositivism." In P. Parrini, W. C. Salmon, and M. H. Salmon (eds.), *Logical Empiricism: Historical and Contemporary Perspectives*. Pittsburgh: University of Pittsburgh Press, 325–346.

Irzik, Gürol, and T. Grünberg. 1995. "Carnap and Kuhn: Arch Enemies or Close Allies?" *British Journal for the Philosophy of Science* 46: 285–307.

Kuhn, Thomas S. 1957. *The Copernican Revolution*. Cambridge, MA: Harvard University Press.

Kuhn, Thomas S. 1970. *The Structure of Scientific Revolutions*. 2nd ed. Chicago: University of Chicago Press.

Kuhn, Thomas. S. 1977. *The Essential Tension*. Chicago: University of Chicago Press.

Kuhn, Thomas, S. 1993. "Afterwords." In P. Horwich (ed.), *World Changes*. Cambridge: The MIT Press, 311–341.

Kuhn, Thomas. S. 1994. "Paradigms of Scientific Evolution." In G. Borradori, *The American Philosopher: Conversations with Quine, Davidson, Putnam, Nozick, Danto, Rorty, Cavell, MacIntyre, and Kuhn*. Chicago: University of Chicago Press, 153–167.

Kuhn, Thomas. S. 2000. "A Discussion with Thomas S. Kuhn." Reprinted in J. Conant and J. Haugeland (eds.), *The Road Since Structure*. 2000. Chicago: University of Chicago Press, 255–323.

Macherey, Pierre. 1978. *A Theory of Literary Production*. Trans. Geoffrey Wall. London: Verso.

Nagel, Ernest. 1949. "The Meaning of Reduction in the Natural Sciences." Reprinted in A. Dando and S. Morgenbesser (eds.), *Philosophy of Science*. 1960. New York: The New American Library, 288–312.

Nagel, Ernest. 1961. *The Structure of Science*. New York: Harcourt, Brace, and World.

Nemeth, Elisabeth. 2007. "Logical Empiricism and the History and Sociology of Science." In A. W. Richardson and T. E. Uebel (eds.), *The Cambridge Companion to Logical Empiricism*. Cambridge: Cambridge University Press, 278–302.

Popper, Karl. 1968. *The Logic of Scientific Discovery*. New York and Evanston: Harper Torchbooks.

Popper, Karl. 1975. *Objective Knowledge.* Oxford: Clarendon Press.

Preston, John. 2008. *Kuhn's The Structure of Scientific Revolutions.* London and New York: Continuum International Publishing Group.

Quine, Willard van Orman. 1951. "Two Dogmas of Empiricism." Reprinted in *From a Logical Point of View.* 1953. Cambridge, MA: Harvard University Press, 20–46.

Reichenbach, Hans. 1938. *Experience and Prediction.* Chicago: The University of Chicago Press.

Reisch, George. 1991. "Did Kuhn Kill Logical Empiricism?" *Philosophy of Science* 58: 264–277.

Reisch, George. 2005. *How the Cold War Transformed Philosophy of Science: To the Icy Slopes of Logic.* Cambridge: Cambridge University Press.

Richardson, Alan. 1998. *Carnap's Construction of the World: The Aufbau and the Emergence of Logical Empiricism.* Cambridge: Cambridge University Press.

Richardson, Alan. 2007. "'That Sort of Everyday Image of Logical Positivism': Thomas Kuhn and the Decline of Logical Empiricist Philosophy of Science" In A. W. Richardson and T. E. Uebel (eds.), *The Cambridge Companion to Logical Empiricism.* Cambridge: Cambridge University Press, 346–369.

Shapere, D. 1964. "The Structure of Scientific Revolutions." *The Philosophical Review* 73: 383–394.

Shapere, D. 1966. "Meaning and Scientific Change." In R. Colodny (ed.), *Mind and Cosmos.* Pittsburgh: University of Pittsburgh Press, 41–85.

Suppe, Frederick. 1977a. "The Search for Philosophic Understanding of Scientific Theories." In Frederic Suppe (ed.), *The Structure of Scientific Theories.* Urbana, Chicago, and London: University of Illinois Press, 1–241.

Suppe, Frederick. 1977b. "Afterword—1977." In Frederic Suppe (ed.), *The Structure of Scientific Theories.* Urbana, Chicago, and London: University of Illinois Press, 617–730.

Wiener, P. 1958. Review of *The Copernican Revolution,* by T. S. Kuhn. *Philosophy of Science* 25: 297–299.

APPENDIX: KUHN'S LETTER

6 October 1995

Dear Dr. Irzik:

A friend just called to my attention to your article "Carnap and Kuhn" in a recent issue of the BJPS. I think it balanced, fair, altogether convincing. As you will imagine, the embarrassment to which I confessed in responding to John Earman is still further increased by your account of the relationship between our views. But I will have to live with that embarrassment, for I cannot complain. My warmest thanks to you and your collaborator for setting the record straight.

Delivery of that appreciation is my purpose in writing, but I will add a few thoughts about the questions with which you close. Part of the problem arose, I think, because of the inevitable human tendency to treat schools of thought as more monolithic that they really are. Carnap had, on your account, moved further from the central position of the Vienna Circle than any of his contemporaries, but he continued to be associated with them by the reading public. In addition, he and I must be seen as moving in opposite directions. He was moving away from his earlier position because of internal difficulties it encountered; I was simply rejecting the tradition (in part because of those same difficulties) and starting again from what I then took to be a very different starting point. That difference did a great deal to disguise the similarity of our views.

Finally, but very speculatively, though Carnap's later works were read (unfortunately not by me), I think very few readers abstracted from them the viewpoint you convincingly describe; they did not, that is, see quite how far he had departed from his earlier views (something he may not have seen fully himself). Certainly, I have talked repeatedly about my views with philosophers who had read Carnap, but until quite recently no one suggested any resemblance. I repeat that I find your account totally convincing, but I also suspect that reading Carnap after Kuhn is not the same experience as reading him before. That is particularly the case because the views you find in him deeply undermine the empiricist tradition. I am not aware of anyone who saw this at the time, and I suggest it would have been too much of a wrench to do so. The case is a little like the distinction between the author of the *Tractatus* and that of the *Philosophical Investigations*. People now see the continuity of views between the two, but my contemporaries thought the later Wittgenstein had denounced his original position.

Sincerely,

Thomas S. Kuhn

P.S. I am sending an identical letter to Teo Grunberg.

3 From Paradigm to Disciplinary Matrix and Exemplar

James A. Marcum

Early in *The Structure of Scientific Revolutions* (SSR), Kuhn defines normal science as "research firmly based upon one or more past scientific achievements, achievements that some particular scientific community acknowledges for a time as supplying the foundation for its further practice" (10). He goes on to identify those achievements rather loosely with the notion of paradigm, which, for Kuhn, is an all-encompassing notion. Based on this notion, Kuhn distinguishes three characteristics of normal science. The first is precision. What Kuhn means by this characteristic is that normal science identifies a "class of facts that the paradigm has shown to be particularly revealing of the nature of things" (25). The next characteristic of normal science is prediction. According to Kuhn, prediction involves the paradigm's capacity "to bring nature and theory into closer and closer agreement" (27). The final characteristic is puzzle solving, which involves the paradigm's articulation or the extension of its scope to include unsolved problems sanctioned by a community of practitioners. For Kuhn, normal science "requires the solution of all sorts of complex instrumental, conceptual, and mathematical puzzles" (36). Thus, the better or more precise and predictive a paradigm, the better or more precise is its puzzle-solving capacity and its ability to predict experimental outcomes. For normal science, then, scientific progress is cumulative or gradual.

On the other hand, according to Kuhn, scientific revolutions are changes or shifts in a scientific community's paradigm; that is, they are "non-cumulative developmental episodes in which an older paradigm is replaced in whole or in part by an incommensurable new one" (91). In other words, scientific progress—in contrast to normal science—is incommensurable or saltatory.[1] Several steps define the specific structure involved in a paradigm shift. The first often encompasses the identification of an anomaly, which is "the recognition that nature has somehow violated the paradigm-induced expectations that govern normal science" (52–53). Given that a paradigm is necessary for practicing science, initially an anomaly may not be problematic for a professional community; that is, the community may elect simply to ignore the anomaly. The next step in a paradigm shift is crisis, which is "a period of pronounced professional insecurity . . . [that] is generated by the persistent failure of the puzzles of normal science to come out as they should" (67–68). The third step is the establishment of extraordinary science. As Kuhn claims, extraordinary science is a transformation of scientific practice such that "the field will no longer look quite the same as it had earlier. Part of its different appearance results simply from the new fixation point of scientific scrutiny" (83). The final step in a paradigm shift is the adoption of a new, incommensurable paradigm. A paradigm shift, according to Kuhn, does not involve simply a community's abandoning an old paradigm without the availability of a replacement paradigm. In other words, a paradigm shift involves competition among incommensurable paradigms and not between the prevailing paradigm and nature. After a revolution, scientists live in a different world from the previous world ruled by the older paradigm.

According to Kuhn, then, scientific advance can be either cumulative, in terms of normal science, or incommensurable, in terms of revolutionary science.[2] Kuhn's representation of scientific advance in these terms elicited criticism—sometimes severe—from a number of prominent historians and philosophers of science (Marcum 2005). One of the main criticisms was Kuhn's notion of normal science. Many philosophers belittled or berated it as an inferior type of science, if science at all, compared to the notion of critical science developed especially by Karl Popper. A problem then arose for Kuhn in defending the distinction between normal and revolutionary scientific progress. For example, in an influential 1964 review of SSR, Dudley Shapere from the University of Chicago bemoaned, "The distinction between paradigms and different articulations of a paradigm, and between scientific revolutions and normal science, is at best a matter of degree" (1964, 388).[3] The problem, as Shapere saw it, was that paradigm is simply too permissive a notion to shoulder the epistemic weight Kuhn places on it. Ultimately, for this critic, Kuhn's distinction or demarcation between normal and revolutionary science depends upon making scientific truth relative.

Kuhn took his critics seriously, realizing the importance of defending the demarcation or distinction between normal and revolutionary science, especially since his philosophy of science depended on it. Without an adequate defense of normal science, particularly in terms of a robust notion of paradigm, revolutionary or Popperian critical science becomes the standard for developing normative notions of science. In this paper, I examine Kuhn's attempts to address critics and to resolve the problems they raise with respect to the demarcation between normal and revolutionary science, which eventually resulted in the addition of the 1969 Postscript to the second edition of SSR. To that end, I begin with Kuhn's 1965 London colloquium paper, followed by the 1967 Swarthmore lecture, the 1969 Urbana symposium paper, and finally the 1969 Postscript. I conclude by discussing whether Kuhn resolves the problems critics raise and by briefly reconstructing a case history from virology supporting the normal-revolutionary science divide.

1965 LONDON COLLOQUIUM

From July 11 to 17, 1965, an international colloquium on the philosophy of science was held at Bedford College located in Regent's Park, London (Lakatos and Musgrave 1970). The British society for the Philosophy of Science and the London School of Economics and Political Science organized the colloquium. On Tuesday, July 13, during the colloquium, a symposium was convened to discuss the growth or progress of scientific knowledge, especially with respect to Kuhn's philosophy of science. The symposium's organizing committee consisted of William Kneale, who served as chair, Imre Lakatos and John Watkins, who served as honorary joint secretaries, and Stephan Körner, Popper, Heinz Post, and John Wisdom. The participants in the symposium, besides Kuhn, included Margaret Masterman, Popper, Stephen Toulmin, Pearce Williams, and Watkins, who at the last moment substituted for Lakatos, and Paul Feyerabend. Popper served as chair of the symposium and moderated a "lively discussion" of the papers presented at the symposium's conclusion.

The papers delivered at the symposium, particularly by Popper, Watkins, and Toulmin, take Kuhn's notion of normal science to task.[4] For these critics, normal science is simply not science at all, especially from a Popperian perspective of science as bold conjectures followed by ruthless testing. In other words, it could not generate knowledge worthy of the designation scientific and so normal science could not lead to advances or growth in *scientific* knowledge. According to Popper, a "normal" scientist is an unfortunate wretch: someone to be pitied and not admired, since he or she is someone "who accepts the ruling dogma of the day; who does not wish to challenge it; and who accepts a new revolutionary theory only if almost everybody else is ready to accept it—if it becomes fashionable by a kind of bandwagon effect" (1970, 52). Such a scientist is poorly educated and trained in that he or she is just indoctrinated but never taught to think critically. Popper calls this scientist an applied scientist, who simply employs the latest techniques to solve a puzzle without any further inquiry, in contrast to the pure scientist, who practices science critically.

Watkins also attacks Kuhn's notion of normal science, claiming that normal science represents periods of "stagnation" in which a scientific discipline becomes mired in its tracks, with no means to proceed forward. Hence, contrary to Kuhn's assertion that normal science provides the backdrop for the detection of anomalies, Watkins argues that it is nothing more than a collection of untested metaphysical dogmas. For Watkins, then, "the condition which Kuhn regards as the normal and proper condition of science is a condition which, if it actually obtained, Popper would regard as *un*scientific, a state of affairs in which critical science had contracted into defensive metaphysics" (1970, 28). Finally, Watkins notes that Kuhn embeds the aberrant notion of science as normal science in a distorted notion of the scientific community. Rather than viewing the scientific community as a

closed society in which dogmas—like paradigms—dictate scientific practice, Watkins advocates the Popperian open society in which such dogmas
are constantly under critical scrutiny.

Toulmin also takes issue with Kuhn's notion of normal science vis-à-
vis Popper's notion of critical science, asserting that "it [critical science]
is always open to scientists to *challenge* the intellectual authority of the
fundamental scheme of concepts within which they are provisionally working—the permanent right to challenge this authority being one of the things
which (as Sir Karl Popper has always insisted) marks off an intellectual
procedure as being 'scientific' at all" (1970, 40). He claims that paradigms
function similarly to R. G. Collingwood's absolute presuppositions, in that
paradigms are nonempirical assumptions about the natural world.[5] Toulmin goes on to challenge Kuhn's absolute distinction between the notions
of normal science and revolutionary science. He argues that the distinction,
similar to that between a presiding government and a political revolution,
is a matter of degree.[6] For Toulmin, absolute discontinuities, or incommensurable paradigms in Kuhn's terms, cannot be the basis for revolutionary
science or progress in science. For example, early twentieth-century physicists, according to Toulmin, gave good reasons for accepting Einstein and
abandoning Newton. Based on his critique, Toulmin concludes, "once we
acknowledge that no conceptual change in science is ever absolute, we are
left only with a sequence of greater and lesser conceptual modifications differing from one another in degree. The distinction element in Kuhn's theory
is thus destroyed" (1970, 45).

In the symposium paper "Logic of Discovery or Psychology of Research,"
Kuhn defends not only the notion of normal science as genuine science,
which advances scientific knowledge cumulatively, but also the distinction between normal and revolutionary science. Kuhn begins his defense
claiming that Popper and his supporters must first take off their Popperian
glasses and put on Kuhnian ones. Only then, asserts Kuhn, could his critics
possibly see their way clear to understanding science and its advancement
as he sees them. Although Kuhn acknowledges similarities between Popper
and himself in terms of science being revolutionary, he criticizes Popper
for failing to see the existence of normal science because of Popper's blind-
ness caused by the brilliance of revolutionary science. For Kuhn, scientists
simply do not have the emotional resources to criticize or test their theories
constantly. Normal science, with its puzzle-solving activity, is necessary in
that it provides a welcome relief from the upheaval associated with revolutionary science. In contradistinction to the Popperian motto "*Revolution in
permanence!*" (Watkins 1970, 28), Kuhn claims revolutions are rare events.
"Sir Karl," argues Kuhn, "has erred by transferring selected characteristics of everyday research to the occasional revolutionary episodes in which
scientific advance is most obvious and by thereafter ignoring the everyday enterprise entirely" (1970a, 19). For Kuhn, "severity of test-criteria is
simply one side of the coin whose other face is a puzzle-solving tradition"

(1970a, 7). Normal science and its steady articulation of a paradigm are generally responsible for the majority of scientific practice and knowledge, at any given period in science's history.

Kuhn ends the paper by defending his notion of normal science on sociological grounds, especially by invoking certain values held by a scientific community.[7] "Knowing what scientists value," argues Kuhn, "we may hope to understand what problems they will undertake and what choices they will make in particular circumstances of conflict" (1970a, 21). One such value includes solving a particularly difficult puzzle, as sanctioned by a specific community of practitioners. Another value involves the quantity of solved puzzles: the more puzzles solved, the better for the community's members solving the puzzles and for the paradigm itself. Other values include the congruence of a theory with related theories, as well as the theory's simplicity and precision.[8] Although these values are important in theory choice or scientific progress, they do not dictate the same theory choice or the same direction for research or progress, simply because community members can utilize the values differently to make individual theory choices or to move in diverse research directions that lead to different lines of progress. However, one value takes precedence over the others—group unanimity. This value, according to Kuhn, is "a paramount value, causing the group to minimize the occasions for conflict and to reunite quickly about a single set of rules for puzzle solving even at the price of subdividing the specialty or excluding a formerly productive member" (1970a, 21). For Kuhn, in agreement with Popper, the psychology of the individual scientist is not singularly determinant in scientific advancement, but, in contrast to Popper, Kuhn claims that shared elements of the community, such as values, are important in such advancement.

Part of the problem for Kuhn's notion of normal science, according to critics, is the notion of paradigm. Paradigm is simply too ambiguous and permissive to carry the epistemic demands Kuhn places on it to justify normal science. Another participant in the symposium provides Kuhn with the wherewithal that eventually helps him to see his way clear to a more precise rendering of paradigm. That participant's name is Margaret Masterman (Braithwaite), a linguist who was influential in founding the Cambridge Language Research Unit in the 1950s. In her paper, Masterman (1970) identifies over twenty ways Kuhn uses paradigm in SSR. She groups these various uses of paradigm into three categories. The first is the metaparadigm, which includes the metaphysical or theoretical basis for practicing science. The next category is the social paradigm, which guides or regulates the activities of a given scientific community. The final category is the construct paradigm, which relates to the practical output of a scientific community, such as solved puzzles. This taxonomic analysis provides an invaluable resource for Kuhn's defense of the notion of normal science and its advances in scientific knowledge.[9] It allows Kuhn eventually to define paradigm precisely as exemplar and to embed it within a larger context of a disciplinary matrix. But, before discussing Kuhn's resolution—as articulated in his later writings, such as the 1969 Postscript—of the problems critics raise, I consider Kuhn's Swarthmore lecture, which provides an invaluable turning point in the development of Kuhn's thought as he struggles to defend normal science and the normal-revolutionary demarcation or divide.

1967 SWARTHMORE LECTURE

Less than two years after the London symposium, Kuhn makes substantial strides toward addressing critics—particularly with the help of Masterman's paradigm taxonomy. In a set of lectures he delivers at Swarthmore College in Pennsylvania, he confronts the problems associated with his philosophy of science—especially with the notion of paradigm. In the second of these lectures—"Paradigms & Theories in Scientific Research"—delivered on February 19, 1967, Kuhn discusses his revision of the paradigm notion. He seeks the audience's indulgence, claiming that his remarks on revising the notion are provisional and tentative. However, Kuhn informs his listeners that he is making progress, "at least in discovering directions in which work must be done" (1967, card 2).[10] The Swarthmore lecture is important because it represents a transition from the initial response Kuhn makes to critics at the 1965 London colloquium, to the mature response, with respect to disciplinary matrix and exemplar, he makes in the 1969 Postscript to SSR's second edition.

Kuhn begins the lecture by framing the problem critics raise concerning scientific advancement vis-à-vis the distinction or divide between normal and revolutionary science. Kuhn reminds his listeners that normal science represents "a tradition-bound activity in which people articulate theories and refine experiments to bring [the] two into closer and closer agreement" and that scientific revolutions pertain to those "episodes which are destructive of one tradition and constructive of a new one. Episodes after which some area of research is quite literally seen differently from the way it had been seen before" (1967, card 3). As is evident from these remarks, he does not abandon the earlier commitment to the normal-revolutionary science divide and its implications for scientific progress. For Kuhn, the demarcation problem is how "to tell in particular cases whether a given development should be described as a normal or a revolutionary advance" (1967, card 3).[11] For the "big" revolutions, such as the Newtonian, Darwinian, or Einsteinian revolutions, there is no problem. But, these revolutions are rare and do not represent the bulk of revolutions in science, which are generally smaller or locally restricted to a particular community of practitioners. Kuhn does acknowledge that for revolutions occurring in the smaller or local community, the distinction between normal and revolutionary advance can be problematic from the perspective of the larger professional and nonprofessional communities in that the revolution may appear as just the accumulation of normal scientific knowledge or it may simply be invisible. Nevertheless, to members of that local professional community, a revolution has occurred. Kuhn's approach to the normal-revolutionary science demarcation problem is two- pronged. He first focuses on the composition of the various communities experiencing small revolutions, and he then examines their commitments in terms of what he calls the professional matrix.

Kuhn begins by informing his audience that "if one wants to determine whether a particular development was revolutionary or not, or even to know quite what this sort of advance is, one must first determine the membership of the group which holds itself responsible for the area of knowledge in which the development occurs" (1967, card 4). As is evident from this statement, he is yoking scientific progress to the composition of a specific professional community. By determining the specific community's response to change in its knowledge or commitments, one can then determine whether it is a normal or revolutionary change. "If anything like a gestalt switch occurs," Kuhn tells listeners, "it is they [members of the professional community] who experience it, and it may be only they. By the same token, they're the bearers of norm[al] sci[ence]" (1967, card 4). Thus, the solution to the demarcation between normal and revolutionary progress rests on identifying a community of practitioners in which a change occurs and then on determining whether the alteration is normal (cumulative) or revolutionary (incommensurable) in the community's commitments.

Kuhn next tackles the problem of how to isolate and investigate the local professional communities, who are responsible for determining if a change represents normal or revolutionary advancement. He bemoans the fact that sociologists have conducted little research to investigate these communities. Although Kuhn warns the audience that he does not base his remarks on any sociological investigation he has conducted, he does provide suggestions on how to examine these communities.[12] Kuhn begins with identifying the larger scientific communities defined by specific disciplinary boundaries. For example, he classifies the various communities with respect to physicists, chemists, biologists, etc. In addition, Kuhn divides these larger classificatory units into major subdivisions. For example, botanists, microbiologists, and zoologists compose the biological science communities. Determining these larger classificatory units of scientific communities, according to Kuhn, is unproblematic. One need simply identify the department in which individual scientists obtained their doctorates, the professional societies to which they belong, the professional journals in which they publish, and the professional meetings which they attend. Moreover, identifying scientific revolutions vis-à-vis these classificatory units is unproblematic, argues Kuhn, since these "major revolutionary episodes cut across groups as large as these" (1967, card 5).

The problem of demarcating between normal and revolutionary scientific advance arises at taxonomic levels of scientific communities smaller than larger divisions and major subdivisions. Moreover, even these smaller groups are sometimes too large to be practically useful for identifying the myriad smaller revolutions that occur in science. Unfortunately, the task necessary for identifying smaller taxonomic units or communities requires new techniques for identifying and studying them. Kuhn informs the audience that this work is being carried out in the discipline of "sociology of science, a field in which substantial work is dreadfully [and] badly needed' (1967, card 7). He then gives the example of two current techniques for getting at these smaller communities. The first is polling individual scientists through questionnaires about their fine-grain professional associations and research. Kuhn informs the audience that he just read a Harvard dissertation that represents a preliminary step along these lines. The second technique is observing footnote citation in the literature, in which members of the relevant community cite one another in specialized journals.

Kuhn concludes the section of his lecture on scientific communities by telling listeners that his goal is to identify the smaller taxonomic groups, since he is convinced that "it is these groups—both in their individuality and in their interrelationships—which must be studied if we're to understand the nature of scientific advance" (1967, card 9). Kuhn is confident that understanding these communities and their commitments will support the distinction between normal and revolutionary science. According to Kuhn, "It's the shared commitments of such a group that make normal science possible," while "it's some element in this body of commitments that is changed at times of revolution" (1967, card 9). By identifying and explicating these commitments, Kuhn believes that he can resolve the demarcation problem that critics raise. For, he contends, these commitments lie at the heart of the nature of science, whether normal or revolutionary, and its remarkable progress in understanding the world.

Kuhn now turns to the professional commitments found in various scientific communities. Earlier in the lecture, Kuhn informed the audience that he wants "to change the way in which the problem [concerning the demarcation between normal and revolutionary science] is discussed in the book [SSR]" (1967, card 4). The change is to focus on the commitments of the professional community. However, he is at a loss for what to call these commitments. "For the cluster of commitments which makes possible a group's research," Kuhn tells the audience, "I need some phrase like the group's professional Weltanschauung, or it's [sic] ideology, or its special matrix of beliefs and values" (1967, card 10). Kuhn settles, only provisionally as we will soon see, on the phrase "professional matrix." This matrix of commitments is composed of various elements, including general statements about nature (theoretical and empirical laws), collective metaphysics, instrumentation and methodology, and the solved problems recognized by the community—the latter, Kuhn tells the audience, was his original intention for the meaning of paradigm—as well as other commitments, which Kuhn does not specify in the lecture.

Kuhn engages the first element of the professional matrix, theoretical and empirical laws, by defining it as "general statements about nature" (1967, card 10). This element represents the formal dimension of the matrix and is the key to any community's commitments. Much of a community's epistemic power and influence, especially in terms of the public eye, arises from these commitments, "for its [*sic*] their presence that permits the application of deductive logic and often of mathematics" (1967, card 12–card 13). Modifications in this element generally signal major scientific revolutions. However, changes also occur at the local level and can represent either a smaller revolutionary advance or normal science development—depending on which community perceives the change. Kuhn utilizes the relationship between current (I) and resistance (R) to illustrate his point.

> The law for heating of a wire by a current, discovered by a number of men including Joule, occurs normally with the conceptual base well prepared. There's no trouble of $H=RI^2$ . . . Ohm's law [$I=V/R$], on the other hand, is denounced, rejected, ignored. Largely because it demands a reconceptualization of the electric circuit before it can be accepted. Notion of resistance in particular participates in revolutionary change (1967, card 12).[13]

What is crucial, according to Kuhn, in these examples of normal and revolutionary change is how a particular community defines these various theoretical and technical terms, how it understands the relationship between them vis-à-vis laws, and how it connects these laws to natural phenomena. Thus, for heating a wire the terms H, R, and I, their relationship with respect to the law $H=RI^2$, and the connection of the law to the natural phenomena were agreed upon and shared by the community, whereas for Ohm's law, as Kuhn points out, R was not.

Kuhn identifies the next element of the professional matrix as "the collective metaphysics of the group. Roughly speaking this consists of the entities and powers which appear in or are used to explain the laws" (1967, card 13). He illustrates this element with entities, such as atoms, for deriving Boyle's law, or forces, such as electrical fields, for deriving Ohm's laws.[14] Kuhn makes two comments on this element. First, metaphysical commitments serve, according to Kuhn, "as guides to research and as parts of theories. Predictions follow from their existence that would otherwise be lacking" (1967, card 13). He illustrates these comments with a historical case study on the nature of light, in which those who explained it in terms of particles looked for pressure exerted by the particles while those who thought it consisted of waves did not. Second, Kuhn remarks, "one need not have agreement on interpretation or even have an interpretation in order to have a scientific community" (1967, card 13). He gives the example of the chemical community at the time of Dalton, who rejected the existence of atoms but still utilized the laws derived from the atomic theory. "Thus,"

Kuhn concludes his comments on this element, "whether or not a particular development is revolutionary for a group may on occasions depend on whether or not a particular metaphysical commitment is an essential element in their matrix of commitments" (1967, card 14). In other words, radical changes in an entity or a force to explain a natural phenomenon often herald a scientific revolution for a particular scientific community.

The third element in the professional matrix composing a scientific community's commitments, Kuhn identifies as instrumental. This element not only includes the obvious technological and methodological dimensions of scientific practice and research, but also involves metaphysical commitments. As Kuhn explains, "The techniques by which we choose to observe and measure the objects of our environment carry with them disguised commitments or expectations about what is and is not in the universe and about the way these things behave" (1967, card 14). In other words, a community's commitments to a specific ontology or to the kinds of entities and forces that populate or compose the natural world often dictate the type of technical instrumentation or protocols used to investigate that world. Kuhn provides an example from astronomy. Historically, astronomers measured time with respect to the uniform motion of stars. But, in the seventeenth century, Newton challenged in terms of his laws of motion the measurement of time based on this uniform motion and with that challenge ensued "a subtle but terribly important shift in the notion of time and of the manner in which it is to be measured" (1967, card 14–card 15).

Kuhn now comes to the final element of the professional matrix—solved problems. He informs his listeners that in SSR, he originally intended the term "paradigm" to denote these problems and his current effort to clarify the notion of paradigm is to recapture that original intention. "These concrete problem solutions, both instrumental and mental," to quote Kuhn, "are the items for which I'd now like to reserve the term paradigm" (1967, card 17). What makes paradigm *qua* solved problems a critical feature of a scientific community's commitments is that these problems are the means by which both science students and practicing scientists connect theories and laws to the natural world itself. Kuhn contrasts his approach to the approach of traditional philosophers of science who presume that the "process of attaching symbolic and verbal expressions to nature is entirely governed by definitions and rules, explicit or implicit" (1967, card 16).

Although Kuhn acknowledges that definitions and rules do function in science, they cannot account completely for how scientists go about setting up problems and solving them. Often, scientists exhibit unanimity vis-à-vis practice even though they may not entirely agree on the meaning of theoretical expressions. According to Kuhn, solved problems function by allowing scientists to see their way to solving new, unsolved problems. In other words, scientists recognize a similarity relationship between the solved and unsolved problems. Important in this process, Kuhn stresses, is "the learned perception of likeness or similarity [that] is prior to and does not imply the existence of a set of criterion [*sic*] which would provide a basis for the judgment of likeness" (1967, card 18). Thus, the set of solved problems *qua* paradigms provides the foundation for normal science practice, and any fundamental change in these problems leads to scientific revolution or paradigm shift.

1969 URBANA SYMPOSIUM

From March 26 to 29, 1969, the University of Illinois in Urbana hosted a symposium to explore the structure of scientific theories, because of the contemporary assault on the traditional view of a scientific theory. The symposium's aim was "to bring together a number of the main proponents and critics of the traditional analysis, proponents of some of the more important alternative analyses, historians of science, and scientists to explore the question 'What is the structure of a scientific theory?'" (Suppe 1974a, vii). Its organizer was Frederick Suppe, along with Alan Danagon acting as co-organizer. The organizers divided the symposium into seven sessions, in which a speaker delivered a plenary paper in a given session, followed by commentary and discussion. The papers, along with commentary and discussion, were eventually published as a book entitled *The Structure of Scientific Theories*. Kuhn delivered a plenary paper during the symposium's sixth session, after which Suppe provided commentary and a number of philosophers, including Peter Achinstein, Sylvain Bromberger, Dudley Shapere, Patrick Suppes, and Hilary Putnam, participated in discussion.[15]

In his lecture, "Second Thoughts on Paradigms," Kuhn acknowledges that part of the SSR's success is the "excessive plasticity" of the paradigm notion. That plasticity is also the root of many of the book's criticisms. Kuhn begins the paper by noting the close association between the notions of paradigm and scientific community. He identifies the association as circular (but not viciously circular, he asserts): "A paradigm is what members of a scientific community, and they alone, share. Conversely, it is their possession of a common paradigm that constitutes a scientific community of a group of otherwise disparate men" (Kuhn 1974a, 460). In order to revise the paradigm notion, Kuhn claims—as he claims in the Swarthmore lecture—that the scientific community needs to be isolated and studied first. He then cites several recent studies by sociologists, who were investigating the social structure of scientific communities.[16] Unfortunately, at the time of the paper, he maintains that the investigations were too preliminary to support a precise definition for a scientific community. Kuhn, therefore, relies on an intuition he feels is shared by scholars to define a scientific community. "A scientific community," asserts Kuhn, "consists . . . of the practitioners of a scientific specialty" (1974a, 461). Members of these communities share goals, educational experience, literature, and conferences, in which there is full communication—although communication across community lines might be difficult at times.

Based on this definition, Kuhn acknowledges that scientific communities are diverse and exist at various hierarchical levels. At the top of the hierarchy is the community composed of all natural scientists. At the next level lower in the hierarchy are the main disciplinary groups, such as physics, chemistry, and biology, in which membership criteria are well defined except at the community's periphery. Major subgroups can also be identified by focusing on a particular or similar technique. These groups consist of organic chemists, high-energy physicists, molecular biologists, and the like. These higher taxonomic groups are easy to identify and characterize in terms of membership. Beginning with the next lower hierarchical level, Kuhn admits that empirical problems arise in terms of identifying and investigating scientific communities. These communities are much smaller, with around one hundred or fewer members. He gives an example of the phage group and claims that to identify this group and its members requires access to preprints, informal communication, summer workshops, etc. This is the smallest unit within the community hierarchy and an individual scientist might be a member of several such communities. "Though it is not yet clear just how far empirical analysis can take us," Kuhn declares in the concluding remarks to this section of the paper, "there is excellent reason to suppose that the scientific enterprise is distributed among and carried forward by communication of this sort" (1974a, 462).

Kuhn then turns his attention to second thoughts on the notion of paradigm. He begins with the following question: "Let me now suppose that we have, by whatever techniques, identified one such community. What shared elements," asks Kuhn, "account for the relatively unproblematic character of professional communication and for the relative unanimity of professional judgment?" (1974a, 462). He notifies the reader that the original answer in SSR is paradigm or paradigm set. However, this usage of paradigm signifies only one of two usages of paradigm in the book. He now calls this first usage "'disciplinary matrix.' 'Disciplinary' because it is the common possession of the practitioners of a professional discipline, 'matrix' because it is composed of ordered elements of various sorts, each requiring further specification" (1974a, 463). The disciplinary matrix is the professional milieu that guides a scientific community's practice of its trade and alterations in any one of its elements may "result in changes of scientific behaviour affecting both the locus of a group's research and its standard of verification" (Kuhn, 1974a, 463). It consists of more than simply a prevailing theory and includes the elements, symbolic generalizations, models, and exemplars. Kuhn then explicates each of these elements of the matrix in turn.

Symbolic generalizations are, writes Kuhn, "the formal, or the readily formalizable, components of the disciplinary matrix" (1974a, 463). For a discipline such as physics, practitioners generally employ generalizations about natural phenomena in symbolic form, such as $F = ma$. However, practitioners can also use words to express generalizations. In biology,

Kuhn cites the example of Rudolph Virchow's dictum "omnis cellula e cellula" or "all cells come from cells." He notes the power of these symbolizations and warns the reader that symbolization is simply an analogy and, if not used carefully, can lead to a distorted image of science. "I believe," concludes Kuhn, "that in several respects we have been victimized by it" (1974a, 465). The symbols of scientific generalizations are more than simply entities that scientists or students instantiate with empirical content from the bottom and then manipulate logically or mathematically at the top. According to Kuhn, these symbols also function as "generalization-sketches . . . whose detailed symbolic expression varies from one application to the next" (1974a, 465). He gives the example of free fall in which $f = ma$ becomes $mg = md^2/sdt^2$. Moreover, empirical content can influence which symbolic generalization scientists or students choose—or even their development of a new one—to solve a particular problem. But, the question arises as to what scientists, and especially students, learn in this process.

To answer the above question, Kuhn introduces the notion of exemplar—the second major usage of paradigm in SSR. But, before examining this element of the disciplinary matrix, another element of the matrix, which Kuhn cites in an earlier listing of these elements, needs to be discussed, even though he warns the reader that he is not going to elaborate on it beyond a few remarks. That element is the notion of model, which supplies a scientific "group with preferred analogies or, when deeply held, with an ontology" (Kuhn 1974a, 463). On the one hand, a model functions heuristically in that it provides a ready means for examining what a natural phenomenon might be like. Kuhn gives the example of an electric circuit, which "may be regarded as a steady-state hydrodynamic system" (1974a, 463). On the other hand, a model may have a metaphysical function, as Kuhn points out in the Swarthmore lecture, in which scientists describe not what a natural phenomenon is like but what it is. He gives the example of a body's heat, which scientists describe as its kinetic energy.

As noted above, Kuhn introduces the notion of exemplar, as the final element of the disciplinary matrix, to answer the question of what students and scientists learn from applying solved problems to unsolved ones. He answers that question by defining the notion as "concrete problem solutions, accepted by the group as, in a quite usual sense, paradigmatic" (Kuhn 1974a, 463). In other words, exemplars provide the template for solving additional problems sanctioned by the prevailing paradigm. Importantly, Kuhn now provides a specific label for what a paradigm is. He goes on to discuss the function of exemplars within the disciplinary matrix, as he did in the Swarthmore lecture, in terms of connecting symbolic expressions to nature and contrasting exemplars to sense-datum language and correspondence rules. He draws upon examples from the history of science to defend the advantage of exemplars over correspondence rules. These examples, Kuhn claims, support his contention that "an acquired ability to see resemblances between apparently disparate problems plays in the sciences a significant part of the role usually attributed to correspondence rules" (1974a, 471).[17] He concludes the lecture by comparing a science student's problem-solving ability to that of a child learning to assemble a puzzle, thereby emphasizing the logical and psychological priority of similarity perception.

1969 POSTSCRIPT

Kuhn informs the reader that the "postscript was first prepared at the suggestion of my onetime student and longtime friend, Dr. Shigeru Nakayama of the University of Tokyo, for inclusion in his Japanese translation of this book" (1970c, 174). Indeed, the only substantial difference between the first and second editions of SSR is the addition of the Postscript. Kuhn makes no systematic revision of the original text for the second edition. Rather, at the time the Postscript represents a first deposit toward "a new version of the book" (Kuhn 1970c, 174). Although Kuhn works almost exclusively for the rest of his career on that project, except for publication in 1978 of *Black-Body Theory and the Quantum Discontinuity, 1894–1912*, he is unable to bring it to fruition in terms of a published book (Marcum 2005). However, the Postscript does represent a mature version of Kuhn's thought, especially with respect to his response to critics. As Kuhn tells the reader, he remains committed to the "fundamentals" of his theory of science. The intent of this essay, as well as the other essays written around this time in response to critics, is to clarify misunderstandings and ambiguities in the first edition of SSR—especially in terms of the paradigm notion, for which Kuhn himself takes partial responsibility.

The first item Kuhn addresses in the Postscript, as he does in the Swarthmore lecture and the Urbana paper, is the relationship between paradigm and community structure. Again, he stresses the importance of isolating and understanding the various communities for revising the notion of paradigm. He gives the same definition verbatim for scientific community that he gave in the Urbana lecture and then discusses the hierarchical social structure, with the "global" community of all natural scientists at the top and the "schools" at the bottom representing the smallest taxonomic social unit, which share a common paradigm and are responsible for the genesis of scientific knowledge. Kuhn contends that understanding these smaller taxonomic schools of scientific practice is imperative to understanding the distinction between normal and revolutionary science, especially since these schools represent "community-based activities. To discover and analyze them," continues Kuhn, "one must unravel the changing community structure of the sciences over time" (1970c, 179–180). The revolutions that occur for the smaller communities are just as significant, Kuhn assures the reader, as the larger global revolutions. Importantly, scientific revolutions—whether global or local—represent changes in a community's commitments, to which he now turns his attention.

Citing Masterman's paradigm taxonomy, Kuhn concedes that the original use of paradigm in SSR was too ambitious and now requires further taxonomic analysis. He also acknowledges that scientists use the notion of theory to account for their practice. Nevertheless, Kuhn protests, "As currently used in philosophy of science, however, 'theory' connotes a structure far more limited in nature and scope than the one required here. Until the term can be freed from its current implications, it will avoid confusion to adopt another" (1970c, 182). To that end, he suggests a "global" use of paradigm as disciplinary matrix, which he defines in almost the same terms used in the Urbana paper. Kuhn lists the components of the matrix—although he warns the reader the list is certainly not complete—to include symbolic generalizations, metaphysical commitments, values, and exemplars. He goes on to note that these components function in unison to guide scientific practice, even though he looks at each component individually for analytic purposes.

As in the Urbana paper, Kuhn begins with symbolic generalizations and defines them in the terms used in that paper. However, he also discusses the relationship of these generalizations to natural laws and, in so doing, bridges both the Swarthmore lecture and the Urbana paper. "These generalizations look like laws of nature," claims Kuhn, "but their function for group members is not often that alone" (1970c, 183). He now includes the insights articulated in the Swarthmore lecture to distinguish two functions for symbols that scientists use to formulate laws. The first is legislative, in which scientists employ symbols not simply to represent and understand the behavior of natural phenomena but also to manipulate them logically and mathematically.[18] The second function is definitional, in which scientists use symbols to define some aspect of a natural phenomenon. Kuhn insists that these definitions can change dramatically over time, in order to explain via natural law a new phenomenon. He gives the example of Ohm's law, which Kuhn utilized in the Swarthmore lecture, and insists that if particular terms, like resistance, "had continued to mean what they had meant before, Ohm's Law could not have been right" (1970c, 183). He concludes by identifying these changes as revolutionary.

Kuhn next discusses the second element of the disciplinary matrix, the metaphysical dimension of a paradigm and the community's commitment to this dimension. In his elaboration on this element in the Postscript, he combines his previous comments from both the Swarthmore lecture and the Urbana paper; that is, he describes this element in terms of metaphysical commitments and models. Importantly, he also defines models not only in terms of preferred analogies but also in terms of acceptable metaphors. "By doing so," Kuhn adds, "they help to determine what will be accepted as an explanation and as a puzzle-solution; conversely, they assist in the determination of the roster of unsolved puzzles and in the evaluation of the importance of each" (1970c, 184). Thus, the metaphysical commitments of a particular scientific community are important in defining what the instrumental practices of its members look like, from sanctioning what puzzles need solving to providing the necessary criteria for determining whether a puzzle solution is acceptable. This element certainly reinforces Kuhn's commitment to the normal-revolutionary science divide, since for normal science this element is stable while for revolutionary science it is changing.[19]

The next element of the disciplinary matrix—values—is added to Kuhn's discussion of the community's commitments, although he initially defends the function of values for scientific practice in the London paper. Values are a vital element in the community's matrix of commitments because they function to provide a sense of community during times of normal science and are particularly important during periods of revolutionary change. Kuhn discusses two important functions of values, although he acknowledges others—such as the social utility of science. The first function pertains to predictions, such as accuracy, quantitativeness, margin of acceptable error, and consistency. The second function involves evaluating theories, especially with respect to puzzle formation and solution, in terms of simplicity, plausibility, and coherence. An important characteristic of values as shared commitment is their application. According to Kuhn, "though values are widely shared by scientists and though commitment to them is both deep and constitutive of science, the application of values is sometimes considerably affected by the features of individual personality and biography that differentiate the members of the group" (1970c, 185). He then defends this position on values from criticisms of subjectivity and irrationality, by noting that shared values are critical not only for determining or regulating group behavior but also in serving science by avoiding turbulent periods of research through distribution of risk vis-à-vis anomalous results.[20]

Kuhn now turns to the final element of the disciplinary matrix, exemplars. As in the Urbana paper, he offers a definition; however, in the present definition, Kuhn avoids using the term "paradigm" and hence he dodges a certain circularity associated with the Urbana definition. According to Kuhn, exemplars are

> the concrete problem-solutions that students encounter from the start of their scientific education, whether in laboratories, on examinations, or at the ends of chapters in science texts. To these shared examples should, however, be added at least some of the technical problem-solutions found in the periodic literature that scientists encounter during their post-educational research careers and that also show them by example how their job is to be done. (1970c, 187)

In other words, exemplars represent a community's collection of solved problems and are generally located in its professional literature, which serves career goals and advancement for its members, and especially in its textbooks, which serve pedagogical purposes for its students. Kuhn concludes by defending the difference between exemplars and correspondence rules, as in his previous essays.[21] And, he emphasizes that through exemplars and the similarity relationships they engender, students and scientists learn substantial things about the natural world in that "nature and words are learned together" (Kuhn 1970c, 191).

CONCLUSION

In response to his London critics, Kuhn (1970b) invokes two types of Thomas Kuhns: the first is the author of SSR and the London colloquium paper, while the other is author of the same works but whose concerns are at times radically opposed to those of the first author—at least according to how Kuhn reads his critics. In the present reconstruction of Kuhn's response to critics, four Kuhns emerge: the London Kuhn (K_L), the Swarthmore Kuhn (K_S), the Urbana Kuhn (K_U), and the Postscript Kuhn (K_P). However, rather than diverging these four Kuhns eventually converge in terms of the continuous development of Kuhn's thought during the 1960s as he responds to critics of SSR, especially with respect to the distinction between normal and revolutionary science and the progress represented by each. The differences among these Kuhns are subtle at times and pronounced at others, as Kuhn labors to articulate more precisely and clearly his view of science and its progress, and thereby to defend himself vis-à-vis critics. A brief summary of the development of Kuhn's defense with respect to these four Kuhns is in order before evaluating the success of his defense.

For K_L, the problems associated with the notion of normal science and the normal-revolutionary science divide come into sharp relief at the London symposium. However, the symposium assists K_L to detect the direction needed to resolve the problems: identify apposite professional communities and clarify the paradigm notion, especially using Masterman's taxonomic analysis. Toward those goals, K_S proposes the professional matrix in the Swarthmore lecture, in which he identifies several critical elements of a scientific community's commitments: general statements about nature (theoretical and empirical laws), collective metaphysics, instrumentation and methodology, and solved problems (Table 3.1). He also proposes that this matrix is important not only for defining the practice of normal science but also for identifying paradigm shifts associated with scientific revolutions. In the Urbana paper, K_U introduces a revised formulation of his response to critics in which he renames the professional matrix "the disciplinary matrix," which consists of symbolic generalization (a broader category than general statements about nature) and models (Table 3.1).[22] In a noteworthy move, he calls the scientific community's repository of solved problems "exemplars." With this reformulation, K_P is confident in his response to critics and thereby enlarges SSR with the Postscript. Specifically, for K_P, the disciplinary matrix includes not only symbolic generalizations and metaphysical commitments and models, but also values as detailed earlier by K_L (Table 3.1). In conclusion, Kuhn defends the notion of normal science and the normal-revolutionary science divide by identifying the communities for whom normal science practice and paradigm shifts are important and by clarifying the paradigm notion in terms of disciplinary matrix and exemplar.

Nevertheless, the question arises: does Kuhn's defense of normal science and the normal-revolutionary divide through clarification of paradigm as disciplinary matrix and exemplar adequately address critics? The answer to that question is both yes and no. On the one hand, Kuhn is able to identify apposite communities for whom normal science and paradigm shifts are important, and he does clarify the paradigm notion in terms of those communities. On the other hand, Kuhn's critics are less than sanguine about the revised paradigm notion. For example, Shapere in a review of SSR's second edition complains that Kuhn's notion of disciplinary matrix does little to clarify the notion of paradigm and even less to justify the "sharp distinction between 'revolutionary' and 'normal' science" (1971, 707). He argues that this current notion cannot bear the epistemic burden Kuhn places on it any better than the original notion of paradigm in SSR's first edition. For Shapere, Kuhn's project implodes under its irrationalism and relativism. Another example is Frederick Suppe's commentary on Kuhn's Urbana paper. Suppe (1974b) expresses appreciation for Kuhn's clarification of the paradigm concept; however, he believes that to defend the normal-revolutionary science divide Kuhn need not populate science or philosophy of science with novel entities like disciplinary matrix and exemplar, or even paradigm, but that he can simply rely on traditional terms like theory.[23] Finally, Alan Musgrave (1971), in an influential review article on SSR's second edition, claims that the notion of disciplinary matrix fails to provide the necessary consensus needed for normal science practice since microcommunities can disagree on metaphysical commitments and still practice normal science. For Musgrave, given the lack of consensus over the fundamentals of scientific practice, "what vanishes is the conception of 'normal science' which was originally attributed to Kuhn" (1971, 289).

Although Musgrave's criticism identifies an important problem with Kuhn's notion of paradigm as disciplinary matrix—especially in terms of metaphysical commitments—for defending the notion of normal science, Kuhn's revision of paradigm with respect to these commitments does provide a means for distinguishing between normal and revolutionary science within smaller communities of practicing scientists. An example from virology supports Kuhn's contention that a community's perception is important in terms of identifying a scientific revolution at a local level, which might appear as normal science at a global level. In 1964, Howard Temin proposed a DNA provirus hypothesis to account for replication of a certain class of RNA viruses (Marcum 2002). Temin's proposal was initially rejected by the virology and molecular biology communities because it violated the well-known and well-accepted central dogma of molecular biology. The dogma states that genetic information flows from DNA to RNA to protein and not in reverse. The DNA provirus hypothesis states that this class of viruses first produces a DNA copy of its RNA genome to insert into the host's genome. The hypothesis was not accepted until Temin reported in 1970 the existence in these viruses of an enzyme catalyzing the reverse transcription of RNA to DNA. At the local level of virologists or molecular biologists, the discovery was revolutionary and led to a new subfield of virology, retrovirology. However, at a global level practicing biologists incorporated the discovery into the central dogma through modification of the dogma and progress appeared cumulative. Kuhn's distinction between normal and revolutionary science helps to reconstruct the history of retrovirology and to furnish an understanding of how the DNA provirus hypothesis became *the* paradigm defining a new field of scientific inquiry.[24]

In conclusion, during the 1960s Kuhn remains committed to the notion of normal science and its cumulative march toward paradigm articulation, that is, its solving of more and more difficult puzzles sanctioned by the community's prevailing set of commitments or disciplinary matrix and thereby adding to the store of exemplars that the community's members use for pedagogical and research purposes. For Kuhn, normal science is what scientists practice the majority of their time, until paradigm articulation begins to fail because of bedeviling anomalies—such as RNA virus replication through a DNA intermediate. If the conditions are right, that is, a new paradigm that solves significant anomalies is available, then a shift from a prevailing paradigm to a new paradigm may ensue leading to a scientific revolution—local or global—in which the way scientists practice their trade changes substantially or even radically. Seen over longer periods, the progress of science is not simply cumulative but also incommensurable; that is, intervals of stasis, or normal developmental growth, are interspersed with punctuated intervals of revolutionary growth. Kuhn believes he is on the right road not only to clarifying more precisely the notion of paradigm but also to defending the notion of normal science and the demarcation between normal and revolutionary science; however, he is not there yet. He spends the rest of his professional career trying to complete the project, especially as it relates to the notion of incommensurability, but death cuts him off before he succeeds (Kuhn 2000).

Table 3.1   Summary of Paradigm Development

K_s Professioanl matrix:	(1) General statements (theoretical and empirical laws)
	(2) Collective metaphysics
	(3) Methodology & instrumentation
	(4) Solved problems
K_u Disciplinary matrix:	(1) Symbolic generalizations
	(2) Models (metaphysics)
	(3) Exemplars
Kp Disciplinary matrix:	(1) Symbolic generalizations
	(2) Metaphysics and models (methodology)
	(3) Values
	(4) Exemplars

NOTES

1. The terms describing scientific progress for normal and revolutionary science, gradual and saltatory, respectively, are evolutionary in nature and reflect its mode. For discussion of the tempo of normal and revolutionary science progress in evolutionary terms, especially as it pertains to the distinction between these two sciences, see Kuukkanen (this volume) and Nickles (this volume).
2. See Chang (this volume) for discussion of the notion of incommensurability, particularly in terms of the chemical revolution.
3. Nickles (this volume) makes a similar point, but he argues that the distinction might be a matter of tempo.
4. The titles of the papers by these three critics provide a synopsis of their thoughts on Kuhn's notion of normal science: Popper's "Normal Science and Its Dangers," Watkins' "Against 'Normal Science'," and Toulmin's "Does the Distinction between Normal and Revolutionary Science Hold Water?"
5. Toulmin is only partially right that Kuhn's paradigms represent Collingwoodian absolute presuppositions. However, Collingwood also identifies relative presuppositions for which empirical evidence can be responsible for changing them. Certainly, Kuhnian paradigms also represent relative presuppositions in that a community changes them based on empirical evidence.
6. Again, see Nickles (this volume) for an alternative approach to the demarcation problem between normal and revolutionary science in Kuhn.
7. "If I were writing my book [SSR] again now," Kuhn writes in response to critics at the London symposium, "I would therefore begin by discussing the community structure of science, and I would not rely exclusively on shared subject matter in doing so" (1970b, 252). He takes up this structure in greater earnestness in subsequent papers.
8. In his essay responding to critics at the London symposium, Kuhn identifies other values that define a community's behavior, such as accuracy, fruitfulness, and scope (1970b, 261–262).
9. Kuhn (1970b) takes up this defense vis-à-vis Masterman's taxonomy in an essay entitled "Reflections on My Critics," added later to the book in which the original papers were published. Unfortunately, the essay takes us too far afield at this point to warrant detailed consideration.
10. Kuhn typed the lectures notes on 5x7 index cards, with pagination at the top right. Thomas S. Kuhn Papers, MC 240, box 3. Massachusetts Institute of Technology, Institute Archives and Special Collections, Cambridge, Massachusetts.
11. Kuhn's demarcation problem is distinct from Popper's demarcation problem, which is how to distinguish between science and nonscience—although from Popper's perspective on Kuhn's notion of normal science, the two demarcation problems are the same.
12. Kuhn's remarks bear the stamp of Pearce Williams (1970), who—in his London symposium paper—chastised both Popperians and Kuhn for not basing their theories of science on empirically derived sociological research.
13. In the Postscript, Kuhn again claims that Ohm's definition of resistance, as well as current, was fundamentally different from previous definitions of the terms. He cites articles by Theodore Brown and Morton Schagrin on this historical episode. According to Schagrin, "the acceptance of Ohm's law required a radical shift in conceptual viewpoint, a revolution in electrical science" (1963, 536). Briefly, rather than the traditional corpuscular view of electricity, Ohm opted for one based on the temperature differential of heat flow. Interestingly, the term "elektrischen Krafte" was rendered as "temperature" in the English translation of Ohm's *Die galvanische Kette, mathematisch bearbeitet* (Shedd and Hershey 1913, 612).

14. The type of metaphysics Kuhn invokes for the second element of the professional matrix is substance metaphysics, which differs from Collingwoodian metaphysics of absolute or relative presuppositions.

15. In a footnote to the discussion section, Kuhn informs the reader that of the three essays he wrote for the 1965 London book project (the essay entitled "Reflections on My Critics"), the 1969 Urbana symposium, and the 1969 Postscript, the essay for the 1969 Urbana symposium was the first written but the last published. "At some of the points where these later essays overlap with my paper in this volume," Kuhn informs the reader, "readers may find evidence that I learn from experience" (1974b, 500).

16. Specifically, Kuhn cites the works of W. O. Hagstrom, D. J. Price and D. de B. Beaver, Diana Crane, and N. C. Mullins, who wrote the Harvard dissertation Kuhn mentions in the Swarthmore lecture.

17. What Kuhn refers to in terms of contrasting paradigms *qua* exemplars to correspondence rules is that exemplars are not simply restricted to the connection of abstract symbolic or theoretical terms with observational terms, as in the case of correspondence rules, but include also training scientists to utilize the similarity relationship between solution of a previously solved problem with a current unsolved one. He gives examples of exercises with an inclined plane or a conical pendulum in which students learn to appropriate what they learn from these exercises to solve closely associated problems. For further discussion on this topic, see Kindi (this volume).

18. Kuhn does not elaborate in the Postscript why he includes the legislative function of symbolic generalizations, but part of the reason most likely involves an expansive notion of paradigm that aims to equip scientists to explore and investigate unknown phenomena.

19. Such changes in a community's metaphysical commitments are only partly responsible for the world changes Kuhn claims occur at times of revolutions. For additional discussion, see Nickles (this volume).

20. Kuhn's defense especially against subjectivity vis-à-vis values is to argue that if all members of a community responded similarly to an anomalous result, little—if any—stability would be possible in terms of normal science activity. What Kuhn (1977) means by subjectivity is not personal bias, which can lead to irrational behavior, but the scientist's individual biography and personality.

21. Besides the Urbana lecture, Kuhn (1970b), in response to critics, also briefly addresses at the London conference the difference between exemplars and rules.

22. An interesting question is why Kuhn shifts the terminology from professional matrix to disciplinary matrix. A possible answer is that Kuhn wants to focus on the theoretical and empirical content of scientific practice in contrast to its social and metaphysical content. He is most likely trying to avoid Shapere's tag of irrationalism.

23. Kuhn (1974b) points out to Suppe that the notion of theory is too limited for accomplishing the task for explaining the behavior of scientists during periods of either normal or revolutionary science—a point he elaborates on further in the Postscript.

24. Peter Vogt (1977), a well-known retrovirologist, states that the hypothesis is the "paradigm" guiding research in the subdiscipline of retrovirology.

REFERENCES

Kuhn, T. S. 1962. *The Structure of Scientific Revolutions*. Chicago: University of Chicago Press.

Kuhn, T. S. 1967. "Paradigms and Theories in Scientific Research." Thomas S. Kuhn Papers, MC 240, box 3, folder 14. Massachusetts Institute of Technology, Institute Archives and Special Collections, Cambridge, Massachusetts.

Kuhn, T. S. 1970a. "Logic of Discovery or Psychology of Research?" In I. Lakatos and A. Musgrave (eds.), *Criticism and the Growth of Knowledge*. Cambridge: Cambridge University Press, 1–23.

Kuhn, T. S. 1970b. "Reflections on My Critics." In I. Lakatos and A. Musgrave (eds.), *Criticism and the Growth of Knowledge*. Cambridge: Cambridge University Press, 231–278.

Kuhn, T. S. 1970c. *The Structure of Scientific Revolutions*. 2nd ed. Chicago: University of Chicago Press.

Kuhn, T. S. 1974a. "Second Thoughts on Paradigms." In F. Suppe (ed.), *The Structure of Scientific Theories*. Urbana, IL: University of Illinois Press, 459–482.

Kuhn, T. S. 1974b. "Discussion." In F. Suppe (ed.), *The Structure of Scientific Theories*. Urbana, IL: University of Illinois Press, 500–517.

Kuhn, T. S. 1977. "Objectivity, Value Judgment, and Theory Choice." In *The Essential Tension: Selected Studies in Scientific Tradition and Change*. Chicago: University of Chicago Press, 320–339.

Kuhn, T. S. 2000. *The Road Since Structure*. Chicago: University of Chicago Press.

Lakatos, I., and A. Musgrave. 1970. "Preface." In I. Lakatos and A. Musgrave (eds.), *Criticism and the Growth of Knowledge*. Cambridge: Cambridge University Press, iv-v.

Marcum, J. A. 2002. "From Heresy to Dogma in Accounts of Opposition to Howard Temin's DNA Provirus Hypothesis." *History and Philosophy of the Life Sciences* 24: 165–192.

Marcum, J. A. 2005. *Thomas Kuhn's Revolution: An Historical Philosophy of Science*. London: Continuum Press.

Masterman, M. 1970. "The Nature of Paradigm." In I. Lakatos and A. Musgrave (eds.), *Criticism and the Growth of Knowledge*. Cambridge: Cambridge University Press, 59–89.

Musgrave, A. E. 1971. "Kuhn's Second Thoughts." *British Journal for the Philosophy of Science* 22: 287–306.

Popper, K. R. 1970. "Normal Science and its Dangers." In I. Lakatos and A. Musgrave (eds.), *Criticism and the Growth of Knowledge*. Cambridge: Cambridge University Press, 51–58.

Schagrin, M. L. 1963. "Resistance to Ohm's Law." *American Journal of Physics* 31: 536–547.

Shapere, D. 1964. "The Structure of Scientific Revolutions." *Philosophical Review* 73: 383–394.

Shapere, D. 1971. "The Paradigm Concept." *Science* 172: 706–709.

Shedd, J. C., and M. D. Hershey. 1913. "The History of Ohm's Law." *Popular Science Monthly* 70: 599–614.

Suppe, F. 1974a. "Preface." In F. Suppe (ed.), *The Structure of Scientific Theories*. Urbana, IL: University of Illinois Press, vii–x.

Suppe, F. 1974b. "Exemplars, Theories, and Disciplinary Matrixes." In F. Suppe (ed.), *The Structure of Scientific Theories*. Urbana, IL: University of Illinois Press, 483–499.

Thomas S. Kuhn Papers, MC 240, box 3. Massachusetts Institute of Technology, Institute Archives and Special Collections, Cambridge, Massachusetts.

Toulmin, S. E. 1970. "Does the Distinction between Normal and Revolutionary Science Hold Water?" In I. Lakatos and A. Musgrave (eds.), *Criticism and the Growth of Knowledge*. Cambridge: Cambridge University Press, 39–47.

Vogt, K. 1997. "Historical Introduction to the General Properties of Retroviruses." In J. M. Coffin, S. H. Hughes, and H. E. Varmus (eds.), *Retroviruses*. New York: Cold Spring Harbor Laboratory Press, 1–25.

Watkins, J. W. N. 1970. "Against 'Normal Science.'" In I. Lakatos and A. Musgrave (eds.), *Criticism and the Growth of Knowledge*. Cambridge: Cambridge University Press, 25–37.

Williams, L. P. 1970. "Normal Science, Scientific Revolutions, and the History of Science." In I. Lakatos and A. Musgrave (eds.), *Criticism and the Growth of Knowledge*. Cambridge: Cambridge University Press, 49–50.

4 Kuhn's Fundamental Insight

Reflection on the "Social Sciences," as a Pedagogical and Philosophical Tool for Thinking Adequately about the Natural Sciences

Rupert Read and Wes Sharrock

The argument that we shall make in this paper is chiefly as follows: To get the central point of Kuhn's work, one need look no further, in *The Structure of Scientific Revolutions* (SSR), than the contrast between the state of sociology and the state of physics, which is literally his founding insight. And one can be very much helped to understand this contrast by understanding Peter Winch—by overcoming the difficulties in understanding him, seeing him too as a "revolutionary" thinker doing something strange and unexpected, something that can be understood and followed provided only that we do not make the mistake of overly assimilating it and him to what we already "know." In sum: to think adequately about natural science, it helps very much to think adequately about "social science"—and *vice versa*.

It is useful to register, from the start, the fact that Kuhn's primary initial motivation was to transform the way in which the *history* of science was written, a change made necessary by the present-minded nature of scientific inquiry.[1] As might be expected, the history produced by the present-minded was poor history, produced only as a subordinate adjunct to other activities in hand, namely those of education in current science; that is, the single largest case of it was the limited amount of more or less "potted" history given *in natural science textbooks*. Though textbooks are important in the training of research scientists, they enable the absorption of the existing literature as a prelude to and complement to scientific research, rather than providing training in how to make (historical or philosophical) investigations. Insofar as the philosophy of science tries to understand change in scientific ideas on the basis of textbook accounts of science's history, then it will be, as Kuhn thought it had been, misled about the way in which science is done, and therefore about the way in which it changes (itself).

Kuhn makes no objection to the present-mindedness of natural scientists from the point of view of doing work in natural science, nor does he even object to the "textbook" history produced for training purposes, but, instead, he objects to leaving the serious writing of history in the hands of those with no great interest in it as such. Teaching natural scientists the history of science—teaching them the history of *their own* natural science (specialism)—isn't necessarily going to give them a better understanding of the science they *now* practice, nor is it particularly likely to make them better practitioners of that science (it is the mastery of current investigative skills that working scientists depend on): "It is by no means clear that a more accurate image of the scientific processes would enhance the research efficiency of physical scientists" (Kuhn 1977, 187n10). It is just this present-mindedness which gives scientific practice (and changes in it) its character, where the transition from one period of work to its successor may involve mutual misunderstandings between those promoting and those resisting change.

We could then sum up Kuhn's task usefully in this way: he aimed to produce, uniquely, an *unscientistic* account of science. His account of science was precisely not the account that practicing present-minded scientists themselves would be likely to produce. But it was an account that aimed *instead* to be *true* to the history of science and to the true nature of science (including crucially, centrally, of *good* (and great) science). True, that is, to the actual scientific practice of science, *not* to self-serving or Whiggish/presentist distortions/formalizations/simplifications of the same.

The problem facing the historian of scientific change is that of preventing one's own (inevitably contemporary) understandings getting in the way of identifying the understandings belonging to the history's subjects. Kuhn's modest proposal (which has had all the controversial consequences) can be summed up in terms of a change in the role that the notion of "rationality" is to play. The more orthodox conception has this question: given what *we* understand, does what they do look rational (to us)? Kuhn's question is different: given what they understand, what do they deem it rational to do? As we shall see, Kuhn's proposal is not meant to be wholly at the expense of the first question, which is the way it is very commonly conceived, but as a means of giving, ultimately, a better, more sensitive and discriminating, answer to that first one—an answer which explains why what looks *obviously* rational to us would in many cases not have even been rational at the time. The problem with "textbook" history of science as Kuhn saw it was that it could not prevent current understandings getting in the way of a better grasp on previous and different ones. Insufficient care was taken with respect to studying previous scientists, in distinguishing what it now looks to us that they must have been doing from what, as far as *they* were concerned, they were up to. Kuhn analogized his problem to that of the anthropologist who ought ideally to "bracket" his or her own standards of the intelligible, acceptable, appropriate, and effective so as to let the indigenous understandings emerge into view[2] (we return to this analogy shortly).

Our book on Kuhn was unusual in including, albeit only in abbreviated form,[3] a serious account of Kuhn's historical studies, to stress that, in Kuhn's own mind, the more generalized account of the structure of scientific revolutions was a means toward the making of better historical studies, as was manifested in his own study of the black-body problem (without explicit dependence upon the terminological apparatus of SSR, though the puzzlement created soon led to a more explicit statement of the dependence—see pages 91–92 of our *Kuhn*).

AN AFFINITY WITH PETER WINCH

As already flagged above, and perhaps unsurprisingly (in light of our and Hutchinson's subsequent book on Peter Winch), we think it helpful at this point to draw a parallel between the moves Kuhn and Winch make, and to say something about the difficulty that these have encountered in relation to the "social sciences." At the time that Kuhn was working on SSR Peter Winch published *The Idea of a Social Science* and then (in 1964) the related paper, "Understanding a Primitive Society," which also generated resistance and controversy that continues into the present. Winch's project was close to Kuhn's, though addressed to the social studies generally, rather than to history or more particularly the study of science's history specifically. Winch's central challenge was that the notion of "understanding" was (and continues to be—cf. Winch 1997) misapplied in the social studies, especially when it is assumed (as it is, at least by default, throughout much of the social studies) that understanding of people's actions only comes in a scientific or theoretical form. One way in which Winch expressed this concern was with respect to the understanding of cultures different to one's own (where those cultures need not belong to alien societies but can be found within the broader culture of one's own society), where the difficulty is less that of needing to adopt more effective methods of inquiry than it is of needing to give up preconceptions that are obscuring the distinctiveness of culturally autonomous standards.

Conventional history and philosophy of science imagine in a certain sense no distance between "them" (the past scientists one is trying to understand) and "us." (Compare Wittgenstein's critique of conventional anthropology: such as his remark that "Frazer cannot imagine a priest who is not basically a present-day English parson with the same stupidity and dullness" (Wittgenstein 1993, 125): Wittgenstein is here bemoaning Frazer's failure of imagination, his failure to conceive of those priests he was studying as anything other than just *a version of the very priests whom he was familiar with from his home society*.) Thus historians of science prior to Kuhn tended to see past scientists as trying to do just the same kind of thing as us, of trying as it were simply to discover the very scientific facts that we have now discovered. We should emphasize that the foregoing is a tendency statement, not an all-embracing generality since Kuhn readily acknowledged historians who were exceptions to this, such as Ludwig Fleck and Alexander Koyré. Philosophers of science such as Hempel, Carnap, or Popper, however, abstracted from the real lived history of science, and thought that timeless and formal rules of scientific method could be established that would determine what the proper choice of theories/of actions was for scientists at any point in history. To actually understand past scientists, we need *to place them at a distance*, sometimes a great distance. (A very useful basis for understanding this is Kuhn's repeated account of how he came to understand Aristotle as a scientist/a physicist—by giving up the assumption he had previously held, that Aristotle must be a physicist in roughly the sense we are familiar with, of that word. See for instance "What Are Scientific Revolutions?" in Kuhn 2000.)

So, it is crucial to be able to see these things *from the right distance*. Not, as it were, from in(de)finitely far away, as if purely observing nonhuman objects; but certainly not in(de)finitely close, either, as if the cultural/intellectual/spatial/temporal differences between us and them could be taken to be inconsequential.[4] Placing of "them" at a greater distance from us *than one is at first inclined to*—this is the hugely innovative step that Kuhn and Winch both take. This step, "ironically," allows one to understand "them" *better*, provided that one is thus first duly warned to bracket overly hasty "false friends" in the "translation," as it were, between "their" schema and ours. (This is not to say that Kuhn and/or Winch have a dogmatic approach to the outcome of exploration into remote practices, that these must conclude in demonstrations that "they" are less like "us" than was initially imagined. These can equally well show that while in some ways they are more distant than a superficial impression suggests, in other previously underplayed respects they may be *more* like us, as when Kuhn suggested that general relativity is more like Aristotelian than Newtonian physics[5])

That is to say of course that, like Winch, Kuhn depended on the assumption that sense or intelligibility is contextual, that expressions are to be understood within the practices in which they have their home, and that he was equally concerned with the risk that inattention to contextual dependence could result in serious misunderstandings. Kuhn's analogy between the historian of science and the anthropologist was meant to point up that the practice of science in earlier periods was located in its own context, one potentially as different from the current scientific context as an alien society can be from the anthropologist's home culture. At the same time, Winch (1964) was pointing out that *anthropology itself* could encounter just the same troubles even when possessed of a very detailed familiarity with the specifics of practices in an alien society. (Thus Winch is very helpful in ensuring that Kuhn's idea actually succeeds in getting us an understanding of the remote, rather than issuing only in the kind of misunderstandings found in Frazer or in Evans-Pritchard.) The real difficulties arise not in getting a grasp on the day-to-day doing of the practice *as such*, but rather at the junction where a clear grasp of the implications of our understanding *them* impacts also upon our understanding *ourselves, and any comparison we might want to make of "us" with "them."* That is to say, in Kuhn's case: the difficulties arise in the attempt to understand the very *nature* of science as a continually (or at least periodically) evolving or changing practice.[6]

A particular difficulty that both Kuhn and Winch saw is then that of the (thoughtless) installation of our own standards of rationality as a general and unilateral yardstick. If they thought that giving up on presupposing *those* standards of rationality for their particular purposes could fix the problem then they surely did not anticipate the extent to which their proposals would create problems that would, in terms of the weight of publication and of opinion, in a certain sense come to rival in assumed difficulty and damage those that they initially intended to solve. The controversy was not only with those who were appalled at Kuhn's and Winch's apparent difficulties with the concept of (universal) rationality, but also with those who, finding their approach inspiring, were then carried away by their enthusiasm.

If the aim was to get those in history and social studies to see the liberating consequences of more scrupulous attention to cases, then this was (sadly) not the practical consequence of either of their proposals. Their readership was not much interested in the recommendation that more careful contextualizing attention be paid to cases, and was much more interested in getting to grips with abstract issues of "principle" concerning the objectivity, intelligibility, and rationality of the natural sciences and/or social studies (as has been plainly shown throughout the enormous amount of commentary on Kuhn, especially by philosophers).

UNDERSTANDING THE CULTURALLY
AND HISTORICALLY REMOTE

At their most modest, Kuhn's and Winch's proposals are simply to give priority to determining just what the subjects of historical (of science) and sociological (etc.) reflections understand themselves to be doing, and to do this *before* bringing in *our* questions about how their practices and constituent understandings match up to our understandings (in respect of cognitive adequacy or rationality). Rather than framing the portrayal of their ways in terms of *our* questions, let a better understanding of what they are up to determine if those questions have relevance—as Kuhn encourages us to think that questions about the rationality, desirability, and efficacy of scientific moves at any given time are ones which belong to the (specific) sciences,[7] not to philosophers and historians. The question "How do they decide when to shift their loyalty from one exemplar to another?" supplants the question "How are we to tell whether they make (or made) the right move in shifting loyalty from one exemplar to another?"

Rather than being treated as reasoned and reasonable proposals that, if we want to compare, evaluate, and criticize someone else's practices, we should first ensure that we are clear about (what) those practices (are), the Kuhn/Winch proposals were seen to carry catastrophic implications for, in the extreme, the very possibility of rational thought. Both Kuhn and Winch are concerned with the problem of understanding other people given the diversity of human practices, and are not concerned with determining in a totalistic way the—cognitive—success of science, "primitive" ritual, or any other genre of practice; but they are overwhelmingly treated as if they are putting out "relativistic" doctrines on just these matters. There is no need to clutter up getting a grip on unfamiliar or distant practice with realist or other metaphysical assumptions and, indeed, involving those in the exercise is likely to be—for the exercise's purpose—wholly counterproductive. Furthermore, as we argue in our two books, Kuhn and Winch had little interest in making *any* metaphysical assertions at all[8] due to their broadly Wittgensteinian inheritance. We would venture to say that, in our view, this is in certain key respects at least pretty obvious, in Winch's case;[9] while in Kuhn's case, compare, as well as our own work, Kindi's *Kuhn and Wittgenstein* (1995).

The following heuristic/propaedeutical/"methodological" observations are being made: (1) that cultures and practices manifest and develop standards and conceptions of rationality, but that these are developed in and through the ways of life that make cultures different; thus (2) it is not the case that one culture—such as ours—is the only one that possesses a standard of rationality; nor (3) is it the case that the standards of rationality developed across different practices are all the same. What we count as rational is a function of what we take to be true of the world at the very general level. In consequence, it is to be expected that the sciences will develop their own standards of rationality (and it is not to be assumed that philosophers' dogmas about those standards are coincident with the actual rationalities of scientific practice), and it is, so to speak, the sciences' business as to what those standards are. It is an objectionable (metaphysical) move to treat the sciences' standard(s) of rationality as if they can readily be applied outside of the context of scientific practice itself, as if they could be used as standards for assessing the rationality of any and every practice, scientific or not, in a way which simply bypasses the question of what the "indigenous" rationality of or appropriate to the practice in question might be. Thus Kuhn and Winch contain clear and well-motivated arguments against any form of scientism, but not in the service of a substantive relativism, nor of any other metaphysical doctrine. For again, what they are *doing* is not what we suppose, and hence our standards may not be *appropriate*. As we mentioned earlier: Kuhn started to see Aristotle as something other than a very—a strangely, surprisingly—bad physicist when he saw how very different Aristotle's fundamental conception of the universe (and thus of his subject) was from our own; similarly, Winch saw the Azande (in his "Understanding a Primitive Society") not as playing our game badly, but as playing *a different game*.

Winch had tried to explain (1964, 315–318), in a response to criticism from Alasdair MacIntyre, the essentially simple point that he wanted to make. Consider the case of people who believe they carry their soul in a stick. Is this a rational thing to do? MacIntyre pronounced that it was plainly not. Winch did not aim to give the contrary verdict, and just as promptly declare it rational. Winch's point was that MacIntyre's contrast short-circuited, comparing the wrong things. MacIntyre compared *their* practice with *our* standards, making no mention of, nor inquiring into, what standards inform *their* practice. Winch wanted to make the *first* question not: does what they do seem rational to us?, but, instead, in what ways is it rational *to them?* (Or, for that matter, *irrational* to them—*if* we are potentially to judge their practice as irrational, then the best place to begin is with any implicit or explicit judgments of irrationality made *by them*.)

MacIntyre's question is asked and answered without familiarity with their lives and ways beyond the fact that, apparently, they can carry their souls around in a stick, presuming of course that "believing that they carry their souls around in a stick" is a sound description of what they do believe. Even if it is in some sense a sound description, relative to what they do and believe, is it really intelligible to us? In what sense? We need to be shown this. What exactly do people who believe they carry their soul around in a stick actually believe? Are we to think of one of "them" as being just like one of "us," as though one of our number—perhaps oneself—suddenly picks up a stick, takes it everywhere he goes, and, if asked what it is for, says "it's got my soul in it." It might well be concluded that he was being pretty irrational, but surely more on the grounds that we do not really understand what he is doing than on the grounds that we have a clear idea of what he is saying when he says, "Be careful with that, it's got my soul in it" or anything like it, and then find *that* irrational. The question is not whether, from a distance, a practice can be assessed as irrational so much as whether, remotely, it can even be identified. Shouldn't we recognize that "their" doing whatever comes under the heading of "carrying one's soul in stick" is probably not much like one of us just setting him- or herself up to do ostensibly the same thing? Among them, seemingly, this is a culturally standardized thing, which puts an individual doing this in a different position to some isolated individual in our own community taking this up, but just being told it is something they conventionally do won't make us much clearer as to what it is they are conventionally doing. The question becomes: What one of *them* does therefore will be intelligible to others, and they won't find the individual's behavior at all puzzling—it will fit comfortably into their relations with each other—so, what *part* will it play then and there?

One is not really in a position (yet) to answer the question "Is that rational?" because one is not even really in a position to understand what is involved in supposing that one can, for example, carry one's soul around in a stick. Winch is not, then, supposing that one takes the activity of "carrying one's soul around in a stick" as a given *and then* establishes what "their" standards of rationality are, to see whether, according to those standards, "carrying one's soul around in a stick" is a rational action. Finding out what their standards are *goes along with*—is partly (co)constitutive of—finding out what sort of practice "carrying one's soul around with a stick" can be. Winch's initial bugbear was the way in which practices that seem, to superficial inspection, very remote from our own can be precipitately rated irrational, whereas Kuhn's initial bugbear was the tendency in history of science to suppose that previous science which, in surface appearance, partly or much resembled our own could be too readily taken to house the same practices or understandings, just in less accurate form. Both Kuhn and Winch suppose that practices are properly understood in context, and that questions about the rationality of some practices are not to be answered—at the very least, are not *to begin with* to be answered—by reference to any supposedly "fully general" standard of rationality, but in relation to the context in which they are actually found[10]—in relation, as it were, (to employ a metaphor that Kuhn became fond of in the years following the publication of *SSR*) to their "ecological niche." The following thought provides the persistent drumbeat to our paper: that identity of what is being done itself comes from context.[11] Again: It is not, that is, a matter of *first* establishing what they are doing and *then*, and separately, establishing what standards of rationality they might apply to decide whether the action is rational or not. Getting to understand what they are doing is part of picking up what their standards of rationality *are*, for those in turn are part of the context. The standards of rationality appropriate to a context, in other words, are in part determinable by reference to what is actually done in that context, for what the context *is* is in part determined by what is actually done in it. It is simply absurd, nonsensical, for philosophers or others to seek to stand wholly "outside" the context which they mean to be "theorizing" and yet still to suppose that they understand what is at stake for those in that context well enough to dictate "correct" standards of rationality/conduct for the context.[12] To attempt to do this is in fact to hover between doing two un-cotenable things.

Winch's main case is that to speak of "scientific rationality" is to remind us that "science" is *one* of our institutions, and that science's standards of rationality are not ones that were pregiven to the development of the institution, but are ones that have formed and developed through the rise of the institution itself. Thus by extension, it is Whiggish and unhelpful to apply such standards straight to whichever culture one is studying. Like Kuhn, Winch does not suppose that science's standard of rationality is a constant of scientific history, but is rather a *part* of it. Thus, historical inquiry is not a matter of figuring out how well work in different periods satisfied one and the same standard of rationality, but of grasping how work in different periods differs in respect of what counts within them as rational.

CHANGING STANDARDS OF SCIENTIFIC ADEQUACY

We've highlighted the affinity between Kuhn's work and Peter Winch's intervention in social studies. This sets the scene for a brief exegesis of the title of our paper: "Kuhn's Fundamental Insight—Reflection on the "Social Sciences," as a Pedagogical and Philosophical Tool for Thinking

Adequately about the Natural Sciences." Our point here is at center just what Kuhn recorded in the Preface to SSR, namely that it was his encounter with living *social* scientists (at the Center for Advanced Studies in the Behavioral Sciences) that actually gave him his theme for his work. The argumentative carry-on—the constant ripping up and starting over, the lack of consensus concerning fundamental questions of method and ontology—among the "social scientists" starkly contrasted with the industrious application of natural science practitioners in mature fields. This made, for Kuhn, the organization of their work the distinctive feature of developed sciences. It was possible for scientific investigators to build on the work of their colleagues, using standardized dependable practices that defined what counted as acceptable scientific results. This was where it made sense to talk about "cumulation" in scientific findings; ergo "normal science."[13] Understanding the possibility of "cumulative knowledge" was a matter of identifying the conditions under which scientists consensually accepted a set of standards embodied in exemplars of proper practice for their area of inquiry. These conditions are plainly absent from large ranges of the social studies, and they obtain only *temporarily* (i.e., *normally*) in the natural sciences. Some natural sciences emerged out of conditions akin to those in the social studies, but most natural sciences have spawned within or spun off from preexisting fields. Even if a developed science does not transmute into a new science, it is in any case apt to undergo occasional (revolutionary) change, between one consensual state and another, where both the substance of the consensus and the identity of the individuals prominent in it will have changed. New exemplars of bona fide work in the field replace earlier exemplars, which—since the exemplars function as *standards* of scientific practice—means that the standards of good or correct science are changing with them, and this may even be the case during periods of normal science as Nickles (this volume) shows.

These "standards of science" are not standards of science in the sense in which many philosophers would perhaps like to have these identified since they are, from the start in Kuhn's work, standards which for the most part have application only in (relatively) specific areas of scientific work, defining, as they do, good scientific practice for that branch of work (over a period of time). Their (relative) specificity is an aspect of their being standards for doing scientific work which, in developed sciences, is in one way or another work within a specialism. The exemplars which identify good science in one field will perforce differ from those in other fields, and likely will alter over time even if the field continues to possess the same "identity." The story of "scientific revolutions" is a tale of the way in which existing standards lose—are deprived of—their local authority and new standards are sooner or later and more or less combatively installed. It is through the transformation of paradigms that the rationality of scientific inquiries is formed.

WHAT DO NATURAL SCIENTISTS KNOW?

To say the least, it is (and this was the central burden of our two books) an irony that both Kuhn and Winch are characteristically construed as showing the very opposite of what they would ostensibly teach. Each gives primacy to getting the understanding of others right, to getting philosophical preconceptions out of the way so as to facilitate seeing the facts in question more clearly; but each is interpreted as if they were doing exactly the opposite, putting philosophical preconceptions in the(ir own) way so as to produce a controversial and challenging picture[14] of those others, one which is surely quite incongruous with those others' own understandings of the practice they are engaged in. Thus, it is as if scientists assumed that they are investigating natural phenomena which are givens to their inquiries but (on the more idealist interpretations of their work) Kuhn and Winch are telling them that they are really only investigating the states of their own minds or conditions which are wholly and purely artifacts—constructions—of their own doings. Kuhn and Winch are accused of portraying others as if they were prisoners of words, entirely enclosed within the circle of their language and, by the same token, completely cut off from anything other than or external to language. If this *were* correct,[15] Kuhn's and Winch's efforts to save those others from unthinking condescension would come at a considerable price to scientists' grasp on the nature of their affairs. It would effectively make them delusional about their own efforts, falsely supposing that they are inspecting and analyzing the properties of independent phenomena, putatively discerning the features of "things as they are" but in fact entirely confined within the space delimited by the conventions of their respective practices (where "within a practice" contrasts with "in contact with the world") . . . Any such turning of the careful "methodological" moment of Kuhn or Winch into some kind of idealist metaphysics exhibits a massive and tragic failure to understand these thinkers.[16] And it exhibits a failure, in particular, to understand—in Kuhn's case, as we shall see in greater detail below—the centrality of his commitment to the view that what any competent scientist who is a master of his or her own research tradition and an active puzzle-solver knows is vast, robust, and richly textured.

It is a double irony that Kuhn and Winch both topicalize the ways in which understanding between people can become problematized when so much of the response to their own work exhibits just that problematic, the controversy "at cross purposes" that Kuhn thinks is characteristic of the transition period between periods of normal science. One probable source of such misunderstanding is the reluctance of many commentators to let go of—even for the sake of argument—assumptions of their own, the better to explore those that Kuhn and Winch alternatively make,[17] with the result that they read from Kuhn and Winch the kind of implications that would follow if their own assumptions were instead being accepted as defining the terms of debate. Though at least some of those assumptions are just what is/are up for question. This is well illustrated by Joel Michell's attempt to put his finger on exactly where and how Kuhn goes wrong—and Michell's type of diagnosis is scarcely unique to him:

> According to such a view [as Kuhn's], we never know how natural systems work, where natural systems are understood as structures existing independently of us and our paradigms. If Kuhn's picture is correct, then we can never know what is really there in the world, existing independently of us and our paradigms.
>
> Kuhn thinks that there is "no theory-independent way to reconstruct phrases like 'really there'" (1970a: 206), and reality, independent of our paradigms, is "ineffable, undescribable, undiscussible" (1991: 12). Friedman (1998) has traced the origins of this sort of view to Kant's (1781/1978) idea that "We have no insight whatsoever into the intrinsic nature of things" (A277). According to Kant, experience is a construction based partly upon the schemata of the mind and categories of cognition, and so there is no way that we can ever know things as they really are. . . . For them all, Kant, the positivists and Kuhn alike, "things-in-themselves" are unknowable. They present variations on the same theme because they start from the same question: what sense can be made of science if "things-in-themselves" are unknowable? (Michell 2000, 644)

Who is included in the "we" that is being denied such knowledge, and what is it that they are being deprived of? Is this "we" they who ask and answer questions that scientists can answer in their capacity as scientific investigators: do you physicists know what the nature of dark matter is?, which might be answered: "At present, no, we don't know much about this. There are theories, etc." Or is the "we," rather, those who want to ask about the "findings" of science generically, independently of any engagement in scientific investigations. They don't then want information about what scientists report themselves as having found in any given case, but (rather) want to ask a question more like whether the things scientists report deserve the title of "findings" or "results" at all. Plainly Michell is reading Kuhn for Kuhn's alleged answer to *that* question, one which he understands Kuhn to give in the negative. If Kuhn is answering in this way then it would seem, as was mentioned above, that accepting Kuhn's view is liable to produce some sort of intellectual catastrophe: "What sense can be made of science if things-in-themselves are unknowable?"

There are various ways in which the issues compacted into this paragraph could be elaborated, some of which respond to the complexities involved in clarifying the issues, but here, for reasons of space, we take the short way with these issues. Set up in Michell's way, the catastrophe would seem to be that the enterprise of science itself has been completely misconceived. Most have supposed that science was telling them about things-in-themselves, and now they are being told that science doesn't tell them anything of the sort. Science points itself at "things-in-themselves" and (allegedly) spectacularly and comprehensively fails to get anywhere near them. The whole point of science is being called into question.

There is no need for these melodramas if one allows that Kuhn is not trying to give an answer—positive or negative—to the question about knowledge in general of "things-in-themselves." Instead, Kuhn is asking whether the notion of "things-in-themselves" has any useful role to play in understanding the way in which working scientists opt for or against new paradigms when realignments are occurring. He need not be understood as denying the term "things-in-themselves" any use at all,[18] but simply excluding the one affiliated with a Kantian idea of "things-in-themselves." Kuhn (arguably, not unlike Kant himself, at his (Kant's) best) is effectively posing this question: is there any real substance to talk of "things-in-themselves" (or of "what is really there in the world"), other than and beyond what current sciences are reporting about the phenomena within their field?[19] Kuhn thinks not; and this is what the above-cited expressions "ineffable, undescribable, undiscussible" in fact indicate, when read carefully, in context. Kuhn's point is similar to Wittgenstein's famous observation that there is no difference between a nothing and a something about which nothing can be said.

The threat of irrationalism that is seen to flow from Kuhn's writing may seem to do so because Kuhn in a certain sense leaves us with absolutely no independent way of saying whether the overthrow of an exemplar—which is always a move to a new exemplar—is for "better" or "worse" . . . but this is only so on the assumption that we should need to make a judgment on this matter on our own behalf and independently of what the scientists themselves think and decide. Kuhn's emphasis on the working scientists' point of view advises that the only question worth asking here is about whether to move from this exemplar to that one, a very specific decision, and one which is inseparable from the practical work of the science and thus from the contemporary technicalities of that area of work. Seeing that, we can see that assessments of "better" and "worse" can't be independent of the scientific deliberations, and reference back to the actual ways in which scientists (variously) respond to a possibly significant anomaly can give us all we really wanted from a general standard of rational progress in science.[20]

Even such a brief explication should make it plain that Kuhn is not arguing that there are "things-in-themselves" and that we can't possibly make contact with them, but that as far as science is concerned there needn't be—and needn't not be—any things-in-themselves of the "Kantian" kind to know about (which is *not* to deny that Kuhn's efforts to detach from "Kantianism" are much more halfhearted than they needed to be).[21] *Give up* that idea of things-in-themselves, don't "play that game," and it is then perfectly possible to say of the phenomena investigated by the sciences that these are things that are really there and that we do know quite a bit about them. Kuhn does not deprive us of certain forms of words, but dissociates their (further) use from surplus philosophical suppositions.

THE RATIONALITY OF WORKING SCIENTISTS

It might, however, seem as though Kuhn leaves matters exposed to skepticism. His very idea of scientific revolutions means, surely, that any "current science" is exposed to the risk of subsequent rejection. What are currently taken for indubitable phenomena may be excluded from future science, and nothing can be done to rule out that possibility. The answer to this is: of course. How else could it be, unless we have or had reason to believe that science has been completed? But Kuhn's more detailed response (manifest in his animus toward Popper) is crucially that this conception of skepticism nevertheless simply *plays no part,* normally, in the day-to-day organization of scientific work. It is not that scientists are in denial about the possibility that radical changes might be made to their science, but that those changes are, from their present position, normally incalculable and almost entirely irrelevant. The possibility of doubt is simply idle throughout periods of normal science.[22]

As we explained above, Kuhn's is no longer a project to articulate a general standard against which the rationality of scientists' decisions may be measured, but rather a project intended to elaborate a description of the rationality of working scientists: the ways in which they decide whether to stick with a going paradigm or shift loyalties to an insurgent one. The change, once more, is from asking "Given what we now understand, was the change of loyalties an unequivocally rational move?" to asking "Given what they understood, what sort of rational options could they discern?"[23] Kuhn's response has two arms. First: historical studies could transform the understanding of how scientific predecessors understood their own tasks and problems, could establish where (so to speak) the contemporary limits of conceivability might lie, and within those limits, why *this* (and not some other) option would seem the optimal one to a working scientist. Thus the famous opening sentence of SSR: "History, if viewed as a repository for more than anecdote or chronology, could produce a decisive change in the image of science by which we are now possessed" (1).

The second arm of Kuhn's response was the insistence that the options facing the scientific group (at moments of crisis) were not, at the time they arose, unequivocal, though they would acquire that appearance in retrospect. One might put the point this way: it is not the scientific consensus that motivates a shift from one exemplar to the next, but the transformation of loyalties that produces the new scientific consensus. For those facing the need to move one way or the other, many of the considerations involved would be incalculable.[24] They would not even be conceivably calculable, that is, in terms of any algorithm.

In an important sense, Kuhn emphasizes the extent to which the main product of—and even the main form of most of the work during—scientific revolutions is: *more scientific work*.[25] As he portrays them, scientific revolutions are about the prospects of rival exemplars, of whether more abundant, challenging, and productive work will result from taking up a new and underdeveloped possibility than will follow from staying with an established and successfully productive one, considerations that, as Nickles (this volume) shows, continue to be present in periods of normal science. Scientists could not know how things would turn out; the adoption and cultivation of the exemplar was the way to find that out. As we have just intimated, Kuhn also emphasizes that those who were to make "the choice" were not in the position of neoclassical rational choosers, not the rational participants of game theory, individuals who could make sound comparisons of the available options, as if each one individually possessed clear and correct mappings of the alternative exemplars based on a sound survey of the respective merits and deficiencies of the exemplars. Kuhn emphasized not only that the choosers could not tell in advance how their options would turn out (by virtue of the very fact of being at the research frontier, at a time of crisis), but that they might well be misinformed about the conditions under which they were making their choices (not least because of misunderstandings following from incommensurability). Periods of revolutionary transition differ from those of normal science in that they are dominated by advocacy-and-investigation rather than just by investigation.

MEASUREMENT AND THE RELATION TO PHENOMENA

We turn finally to Kuhn's treatment of the function of measurement in the natural sciences, which he understands to be chiefly the practice of reporting phenomena in numerical terms. Looking at this can do further work in dispelling the idea that Kuhn's account of scientific revolutions leaves scientists cut off from the natural or physical world (while simultaneously guarding against crude scientistic aspirations on the part of "social scientists"). In continuing his transformation in the problematic of philosophy of science toward taking the point of view of the working scientist into account focally, Kuhn displaces the abstract question of whether the achievements of science match "how things really are" with one which asks how working scientists satisfy themselves that there is a fit between how they figure things are and how those things are.[26]

His treatment of this presupposes his fundamental argument from SSR that it is useless to attempt comparative global judgments between bodies of work conducted under different paradigms so as to say that one comprehensively "corresponds with the facts" and another does not. This is *not* because Kuhn has or necessarily need have any philosophical anxieties whatsoever about either the phrases "corresponds with the facts" or the use of the words "true" and "false" in their vernacular sense and independent of philosophical theories. It is more by way of observing that, within the body of statements generated by or generatable from any given scientific approach, there will be some which are true and some which are not, some of which correspond with the (best) data and some of which do not. The relationship of this observation to Kuhn's account of scientific *revolutions* is too complicated to go into here,[27] save to say that such transitions perforce involve the scientific community in more or less global judgments, but ones which must not be imagined as involving the individual scientists opting between the possibilities on offer as themselves making point-for-point comparisons across the board of each paradigm's suppositions and results. The whole idea of "anomaly" enacts that point: it is, as such, a routine and unremarkable fact that many of a paradigm's own research results are—throughout the life of the paradigm—out of line with the expectations derivable from it, but this makes the lack of fit at numerous points no more than (in Harold Garfinkel's terms) a normal trouble of normal scientific work. Kuhn's working assumption is that the nature of nature is more complex than is comprehensively rendered under any exemplar, and that its manifest and recordable behavior invariably overflows the limits of the exemplar's "conceptual schemes." He also assumes that this is something scientists simply live with in many respects and for much of the time. Anomalies matter in cases where, for any of diverse reasons,[28] they acquire strategic significance relative to the exemplar—for example, does the persistence (not just the occurrence) of an anomaly eventually make it a downright embarrassment for the paradigm?

In those terms, it is clear that working scientists are aware that there are respects in which the exemplars that they follow do not ubiquitously and uniformly fit with the results of the research they inspire, but in his 1961/1977 paper on quantification and measurement, Kuhn asks about cases in which there is a fit between the exemplar and the results of research: what counts as "a fit"? In short, this is not a matter which is determined by any general or transcendental standard since it is, effectively, that of the abstractly indeterminate "good enough." Working scientists develop consensus among themselves as to when the results of measurement are good enough to count as being "in agreement" with their expectations, and what is agreeable varies between one area and another. Kuhn (naturally) declines the "textbook" picture of the relationships between, for example, theory and research on the grounds that (1) the textbook picture gives the false impression that the role of research results is to confirm theories, whereas it is often commitment to the correctness of a theory which determines which numerical results scientists will *accept* as measurements; and (2) the picture gives in any case a misleading picture of how scientists get to the settled, tidily theory-fitting measurements reproduced in textbooks. The textbook picture misleads in suggesting that "new laws of nature are . . . discovered simply by inspecting the results of measurement made without advance knowledge of these laws" but this is "very seldom" the case:

Because most scientific laws have so few quantitative points of contact with nature, because investigations of those contact points usually demand such laborious instrumentation and approximation, and because nature itself needs to be forced to yield the appropriate results, the route from theory or law to measurement can almost never be travelled backward. (Kuhn 1977, 197)

Kuhn has no difficulty in writing of scientific inquiry as an interaction with "nature" through measurement practices:

If what I have said is right, nature undoubtedly responds to the theoretical predispositions with which she is approached by the measuring scientist. But that is not to say either that nature will respond to any theory at all or that she will ever respond very much. (Kuhn 1977, 200)

Relevant "responses of nature" occur only under restricted conditions and their production is often a matter of considerable difficulty and protracted improvement in measurement practice: it is anything but easy, and may even require "occasional genius" to figure out how to address nature and to develop effective means of implementing the attempt in such a way that the undertaking's output will count as an answer to a worthwhile scientific question. The working scientist's concern is

with theories that seem to fit what is known about nature, and all these theories, however different their structure, will necessarily seem to yield very similar predictive results. If they can be distinguished at all by measurements, those measurements will usually strain the limits of existing experimental techniques. Furthermore, within the limits imposed by those techniques, the numerical differences at issue will very often prove to be quite small. Only under these conditions and within these limits can one expect nature to respond to preconception. On the other hand, these conditions and limits are just the ones typical in the historical situation. (Kuhn 1977, 201)

If some of Kuhn's critics have regarded the role he assigns to exemplars as meaning that scientists are denied access to the nature of "things-in-themselves," then the preceding quotes should justly falsify such an impression. They emphasize the extent to which the existence of exemplars contributes to the viability of scientific investigations as transactions with nature through highlighting the role of quantification in articulating the relation between nature itself and current ideas about and current practices for manipulating it for scientific purposes. As Kuhn portrays their role, exemplars potentiate the possibility that "the results of measurement can be expected to make [scientific] sense" (1977, 200), the conditions under which "laboriously" developed artifacts can provoke and display reactions from nature in forms that can relevantly be counted as answering the questions that scientific investigators put, some of which will provide, in the "historical situation," grounds for concluding that the prevailing exemplar is on broadly the right lines about the nature of the phenomena it understands itself to be addressing. It is only at particular junctures in the development of a science that it becomes intelligible to ask, and possible to answer (though *not* necessarily decisively), the specific phenomena-directed questions that working scientists ask.

Kuhn thus very much *acknowledges* the importance of quantification in the modern physical sciences, but is suggesting that this importance may be overstated (especially in philosophical or methodological pictures of the nature of science) *if* quantification is imagined as an end in itself, especially if the progress of science is identified with the extension of measurement. Kuhn has argued (see for instance his account of *The Copernican Revolution*, 1957) that many major innovations in scientific ideas are made in advance of any possibility of measurement and may facilitate the development of that measurement. In his view, the capacity for effective measurement is relative to the prior cumulation of *qualitative understandings of phenomena*.

If Kuhn's account of the natural sciences originated in his realization of how they differed, as working environments, from the world of the social studies, it is at this point that his arguments reprise *for* the social sciences. At the time of SSR's writing, the social studies were very much (and still in a number of cases to some very considerable extent are) in thrall to a picture of science as defined by measurement, being no less aware than Kuhn that their disputatious, programmatic, inconclusive, and methodologically loose practices were not achieving cumulating knowledge of the sort that was to be found in the physical sciences. The "failure" to progress in that way was commonly ascribed to the advanced numeracy of the natural sciences, to the use of mathematics and the capacity to quantify—which "the social sciences" had "yet" to match. Though Kuhn's lesson about the formation and development of the natural sciences was that this was an emergent and unplanned development, this was not then recognized in social studies, where the response to the "gap" between social studies and physical science was to suppose that this needed to be overcome, and that the way to do this was to develop a "mathematical sociology" (or a "thoroughly mathematicized economics," etc.) and thus to attack "the measurement problem." Kuhn's discussion of quantification indicated that pursuit of measurement for measurement's sake was unlikely to be effective and that the conditions for following the precedents of the physical sciences were not auspicious, given that the social studies had arguably simply not accumulated the consolidated *qualitative* understanding that goes before scientifically functional measurement practices. Since that time, whether from Kuhn and/or from other sources, many in the social studies have come to the view that solving "the measurement problem" isn't the key to—or at least, isn't a currently *viable* means of—transforming their disciplinary prospects, if they are not now skeptical of the possibility or need of making, for example, sociology "more scientific," let alone "more quantitative." Even so, the ghost has certainly not been entirely given up and there are still those (including, worryingly, increasingly many politicians and administrators/bureaucrats/ managers) who imagine that there is something privileged about researches which promise and produce numerical outputs, that these embody a more "scientific" spirit, and embody the genuine attempts to upgrade the social studies into social *sciences*. Those who are drawn to this route may be unaware of the force of Kuhn's general arguments (including those of SSR as well as the "quantification" paper) that there is no reason to expect that calculated efforts to adopt the forms of "scientific inquiry" are any more likely to make over a field without exemplars into one that is exemplar-based than other, relatively "aimless", exercises (bearing in mind that there is no necessity that such a makeover will *ever* take place, and some good reasons to think it will not.[29]).

CONCLUSION

The various remarks about the "philosophical sociology" of science offered in the present paper,[30] and the totality of the Kuhnian "reminders" we have endeavored herein to furnish, disclose a key respect in which our title tells the story that needs telling. Fear and denial of the plainest truism—that science itself is a social process, an activity undertaken by a community—lies at the root of much of the overdetermined antagonism toward Kuhn; while overplaying of that truism by some sociological theorists and "social theorists of science" (who have (over)interpreted Kuhn) has been responsible for some of the same. It was in large part to show how ill-motivated most of that antagonism is that we wrote our book *Kuhn: Philosopher of Scientific Revolution*, the first full-length effort in the English-speaking philosophical world to undertake a more or less Wittgensteinian and genuinely social and historical understanding of Kuhn's philosophy.[31]

Since the publication of that book of ours nearly a decade ago, new studies have continued the project of getting Kuhn right. In terms of those which recently we have been particularly impressed by, we should like to mention in particular Bojana Mladenovic's *Kuhn's Legacy: Epistemology and Metaphilosophy in Kuhn's Mature Thought*,[32] and also some of Jouni Kuukkanen's work, such as for example his "Rereading Kuhn" (Kuukkanen 2009). Much of what such authors say we greatly applaud; it not infrequently extends, details, corrects, or improves upon what we have done.

But we believe that the single best hermeneutic for Kuhn remains that which we stressed at the opening of our book's exegesis of Kuhn,[33] and which we have stressed even more strongly here. That hermeneutic, which gives the present paper its title, is the huge clue that Kuhn himself gives, in the Preface to SSR, as he tells his readers of how he first came to the central insight of his book:

The final stage in the development of [SSR] began with an invitation to spend the year 1958–9 at the Center for Advanced Studies in the Behavioral Sciences. . . . [S]pending the year in a community composed predominantly of social scientists confronted me with unanticipated problems about the differences between such communities and those of the natural scientists among whom I had been trained. Particularly, I was struck by the number and extent of the overt disagreements between social scientists about the nature of legitimate scientific problems and methods. Both history and acquaintance made me doubt that practitioners of the natural sciences possess firmer or more permanent answers to such questions than their colleagues in social science. Yet, somehow, the practice of astronomy, physics, chemistry or biology normally fails to evoke the controversies over fundamentals that today often seem endemic among psychologists or sociologists. Attempting to discover the source of that difference led me to recognise the role in scientific research of what I have since called "paradigms". These I take to be universally recognized scientific achievements that for a time provide model problems and solutions to a community of practitioners. *Once that piece of my puzzle fell into place, a draft of [SSR] emerged rapidly.* (SSR, vii–viii; emphasis added)

To get the central point of Kuhn's work, then, one need look no further than the contrast between the state of sociology and the state of physics.[34] And understanding Winch—overcoming the difficulties in understanding him, seeing him too "at the right distance" (and, through him and Kuhn, seeing distant allegedly or actually technological and scientific practices at the right distance), seeing him too as a revolutionary thinker doing something strange and unexpected, something that can be understood and followed provided only that we do not make the mistake of overly assimilating it and him to what we already "know"—can very much help one to understand this contrast.[35]

To think adequately about social science, it helps very much to think adequately about natural science—and vice versa.[36]

NOTES

1. One initial indicator of why this was important, of course, has to do with Kuhn's well-motivated suspicion of most philosophers of science: a suspicion of their not doing history/being historical enough. Nuanced historical research is likely to reveal the contextual aspects of science in a way that an abstract philosophical reflection cannot. We elaborate on this below.
2. He also, like Wittgenstein, analogizes his problem and method to that of the practicing (as opposed to theorizing) psychoanalyst or therapist: "I think that a lot of what I started doing as a historian, or the level of my ability to do it—"to climb into other people's heads", is a phrase I used then and now—came out of my experience in psychoanalysis . . . I think that what gets forgotten is that there is a craft, hands-on aspect [to therapy/analysis] that

I know no other route to, and that is intellectually of vast interest" (Kuhn 2000, 280).

3. The word limit imposed on us by the good people at Polity cramped our style somewhat in this regard—we wanted to write (and did indeed write) twice as much on Kuhn's historical studies as we were in the end allowed to publish.

4. These metaphors of distance should *not* of course be taken too literally. The metaphor of "infinitely far away" here is intended to bring into prominence the objectual and plainly independent nature of (for instance) astronomical objects, as an analogue for human/social beings; but of course atoms or billiard balls could serve equally well. While the metaphor of "infinitely close" is meant to bring into prominence the lack of interest in the specificities and/ or varieties of cultural and other contexts that are characteristic of overly universalist or blanket-rationalist approaches.

5. And as for instance when Winch suggests that we try thinking of (a substantial portion of) Zande magic as actually quite like Christian prayer (on an uncondescending, broadly Wittgensteinian interpretation thereof), as opposed to like technology and science. The question is: of finding the apposite *object of comparison*, to use Wittgenstein's important term. (If we had space and time, we would like to elaborate this point about underplayed respects in which "they" resemble us and about finding the most illuminating object(s) of comparison with a (Wittgensteinian) dwelling on the closely connected centrality of *aspects* to a full understanding of these matters. Similarity is always under an aspect; aspects are to some extent flexible, dependent upon experience and others' leanings; and we should mention also the idea, central to Winch (if implicit), that processes of interpretation and understanding require selection of aspects on the part of interpreters/understanders/describers. In other words, the concept of aspect most helpfully allows one to look at historical or anthropological interpretation, understanding, and description as a form of dialogue in which the interpreter/understander/describer must make consequential choices. For more on some of this material, see Chapter 1 of our and Hutchinson's (2008), and also much of Hutchinson and Read's recent work on Wittgenstein.)

6. Cf. our *Kuhn* (217n7, 291n23).

7. In this absolutely crucial regard, Kuhn is therefore of course, and contrary to his reputation, the *least* philosophically-aggrandizing of all philosophers of science, and the most "pro-science"! Compared to him, Carnap, Popper, Lakatos, Laudan etc. are all philosophical imperialists, who think that rationality is for philosophers, not scientists themselves, to adjudicate. We develop this point at some length, below.

8. We recognize this will be perceived by many as a contentious claim, and that it contradicts the surface appearances of Kuhn's (and Winch's) writings at various moments. Andersen (this volume) certainly takes a different view from us on this claim / on this matter.

9. If the reader doesn't find it relatively obvious, then we recommend consulting the closing pages of Chapter 1 of our and Hutchinson's (2008).

10. It is important to be aware that, contrary to the impression that can be given by essays such as "The trouble with the historical philosophy of science" (collected in Kuhn 2000), written to slough off some of his over-keen "fans," Kuhn actually if anything *deepened* his commitment to this rationality-contextualism, as the years went by; e.g., "on the developmental view, scientific knowledge claims are necessarily evaluated from a moving, historically situated, Archimedean platform" (Kuhn 2000, 95). Also his idea of science and scientists as adapting and forming their niches, including tools and practice, supports such emphasis on context—for discussion, see our *Kuhn* (188ff.).

At times, Kuhn nevertheless stresses a fairly standard interparadigmatic set of epistemic values (see especially "Objectivity, Value-Judgement and Theory Choice," in *The Essential Tension*). Kuhn even writes at one point that "accuracy, precision, scope, simplicity, fruitfulness, consistency, and so on, simply *are* the criteria which puzzle solvers must weigh," and are "necessarily permanent, for abandoning them would be abandoning science together with the knowledge which scientific development brings" (Kuhn 2000, 251–252). But none of this counts against incommensurability nor against contextuality—because even these virtually ubiquitous scientific virtues are of course open to interpretation, and moreover, they somehow (as Kuhn mentions), imponderably, are to be *weighed*, one against the other, by practicing scientists. For more on this matter, see Read's "The Incommensurability of Scientific Values," section 1.4 in his *Wittgenstein among the Sciences* (2012).

11. And this thought springs, of course, from deeper roots in Wittgenstein and Harold Garfinkel, more recently from and in Charles Travis and Lars Hertzberg—and, we would venture to add, from common-sense itself.

12. For more on this, see Chapter 3 on ethnomethodology in our and Hutchinson's *There Is No Such Thing as a Social Science* (2008). What is at stake here is partly what ethnomethodologists call "accounting," and its *internal relation* to the practices which are being "accounted" for—see for instance the Introduction of Michael Lynch's *Scientific Practice and Ordinary Action* (1993) for explication. (One might compare (and *contrast*) here Bruno Latour (of *Science in Action* (1987)) including on the Azande. For while Latour (at 186–193) makes a point about researchers' need to discover, rather than impose, what Zande standards of rationality involve, he does in the end simply "translate" their speech into "our" language. He sees the need for appreciating local standards of rationality but fails to see what we emphasize in our text: namely, that contextualizing standards and canons of reasoning and arguing requires contextualizing *everything else* too, ontology included.)

13. See the "Normal Science" chapter of our *Kuhn*.

14. Or perhaps "nonpicture," if relativism acts as an impairment to our understanding the other. For discussion, see 1.1–3 of Read's *Wittgenstein among the Sciences* (2012).

15. We argue of course that it is not, in Chapter 2 of our and Hutchinson's (2008).

16. Although one ought to take care also not to travesty idealism, which is/was frequently a lot subtler than its reductive critics have taken it to be. "Idealism" is now a stock figure, routinely ridiculed or dismissed—idealism has become a victim of a Whiggish history, told by the victors, within philosophy itself, in exactly the kind of manner that Kuhn warned of, in the sciences. In fact, the idealists' vision of "internal relations" was a deeply serious one, which has justly influenced Wittgenstein and Winch to some degree, and which it is unclear whether Russell or Moore ever really genuinely refuted; and Winch was also of course influenced directly by Collingwood's deep (albeit not unproblematic) understanding of the philosophy of history. Just consider Winch's book's title, which so clearly pays Collingwood a deep homage.

17. In other words, it is unfortunate, albeit not in the end very surprising, that commentators didn't at least try out the very hermeneutic that Kuhn and Winch recommend, in an effort to understand the strange things that Kuhn and Winch appear to be saying. (Cf. Read's 2003 paper, "Kuhn: le Wittgenstein des sciences?", for one effort to do just this, vis-à-vis the strangest things that Kuhn ever said; or see section 1.1 of Read's *Wittgenstein among the Sciences* (2012).)

18. For this would be an unnecessarily excessive, an implausible, and in any case an ineffectual claim, more akin to Rortian or Carnapian method than to anything genuinely Wittgensteinian.
19. And here, Kuhn (and Wittgenstein too) is strikingly close to an increasingly strong strand in contemporary philosophical naturalism.
20. This is perhaps a natural moment to underline the truths/truisms that Kuhn very much believed (1) in scientific progress, (2) in there being "no going back" after revolutions (a phenomenon he termed "irreversibility"), and (3) that what is normally meant by "relativism" about science was a philosophically, historically, and sociologically hopeless bad joke. See the final section of SSR, "Progress through Revolutions"; the discussion of Kuhn and Winch in Chapter 1 of our and Hutchinson's (2008); and also the whole of our *Kuhn*, but especially Chapters 3 and 4.
21. Compare the "Incommensurability 2" chapter of our *Kuhn*. We urged in that book that Kuhn be read as closer to Wittgenstein than (as Hoyningen-Huene would have it) to Kant—but we also admitted that there are closenesses to Kant in Kuhn, closenesses that continue to cause problems for him.
22. For an acute discussion of the meaning of the idleness of such doubts, cf. Wittgenstein's *On Certainty*, passim.
23. It is salutary here to compare (and contrast) Kuhn to Feyerabend. The latter often argued that scientific change is not always rational even in the light of working scientists of the time when the change was made; that scientists typically theorize not only ahead of conclusive evidence but despite evidence that seems conclusive against their views; and that science is none the worse for that—quite the contrary.
24. We mean this in roughly the same sense in fact in which Wittgenstein speaks of "imponderable evidence" (*Philosophical Investigations* II, xi, 228).
25. For an "exemplar" of this, see pages 80–93 of our *Kuhn*, our discussion of the black-body problem and of how Kuhn revisioned most of the seeming quantum revolution as in the main "just" more normal science (such that Planck turns out significantly less revolutionary than had previously been thought). Kuhn makes a similar move with regard to Copernicus, the "last Ptolemaic astronomer."
26. Here once more there is a clear link with ethnomethodology, this time with the "ethnomethodological studies of work" line of work. In ethnomethodological studies of work, one looks to "lay experts" rather than to social scientists to help one understand one's chosen object of study—for example, to coroners rather than to sociologists, to learn about suicide. Similarly, Kuhn's fundamental impulse is to trust *the scientists themselves,* rather than to look to "meta-" studiers such as positivist or Popperian—or any—philosophers.
27. See the two "Incommensurability" chapters of our *Kuhn*.
28. Explored in some detail in SSR and in Kuhn's historical studies.
29. On which, see our and Hutchinson's (2008), where we detail what (as we have briefly indicated here) is (for good reason) lacking typically in "the social sciences" (and why their accumulation of reams of "data" is not genuinely analogous to data collection in the natural sciences): namely, qualitative understanding abstracting from participants' own understandings; and where we explain in much more detail the unavoidable import of just such understandings-in-action, which impose a limit on the very possibility of approaching the social world scientifically.
30. It is crucial to note here that Kuhn is developing history (his historical case studies are the key *evidential support* he gives for the ideas of his that we have just sketched) and philosophy much more than he is developing sociology proper. See pages 109–113 of our *Kuhn*.

31. Kindi's Greek-language work, *Kuhn and Wittgenstein: Philosophical Investigation of the Structure of Scientific Revolutions* (1995) of course preceded our book by several years.
32. Forthcoming from Columbia University Press.
33. At page 26.
34. Though, once more, and crucially: *without* falling into the delusion that sociology itself ought to *become* like physics—see the argument of our and Hutchinson's *There Is No Such Thing as a Social Science.*
35. We suspect strongly that one reason why so many people have been unwilling to comprehend Kuhn's revolution is because it is (when properly understood) so "unexciting"—lacking in social *caché*—compared to the excitement involved in asserting that science is really just mob rule, or that scientists do not really make contact with the world, or that scientists are basically a kind of literary critic, etc. (And somewhat similar considerations apply to the standard "overinterpretation" of Winch.)
36. *Thanks* for helpful prior readings of this material to the editors, and to Craig Taylor, Angus Ross, Ruth Makoff, and other colleagues.

REFERENCES

Andersen, H. This volume. "Scientific Concepts and Conceptual Change."
Hutchinson, P., Read, R. and Sharrock., W. 2008. *There is No Such Thing as a Social Science.* Aldershot: Ashgate.
Kindi, V. 1995. "Kuhn's *The Structure of Scientific Revolutions* Revisited." *Journal for General Philosophy of Science* 26: 75–92.
Kindi, V. 1995. *Kuhn and Wittgenstein: Philosophical Investigation of the Structure of Scientific Revolutions.* Athens: Smili Editions.
Kuhn, T. S. 1957. *The Copernican Revolution: Planetary Astronomy in the Development of Western Thought.* Harvard University Press.
Kuhn, T.S. 1962. *The Structure of Scientific Revolutions.* Chicago: University of Chicago Press.
Kuhn, T. S. 1977. "The Function of Measurement in Modern Physical Science." In *The Essential Tension.* Chicago: University of Chicago Press, 170–224 (Originally published in *Isis* 52, No. 2 (June 1961): 161–193).
Kuhn, T. S. 2000. *The Road Since Structure.* Ed. J. Conant and J. Haugeland. Chicago: Chicago University Press.
Kuukkanen, J. This volume. "Evolution as Revolution. The Concept of Evolution in Kuhn's Philosophy."
Kuukkanen, J. 2009. "Rereading Kuhn." *International Studies in the Philosophy of Science* 23: 217–224.
Latour, B. 1987. *Science in Action.* Oxford: Blackwell.
Lynch, M. 1993. *Scientific Practice and Ordinary Action.* Cambridge: Cambridge University Press.
Michell, J. 2000. "Normal Science, Pathological Science and Psychometrics." *Theory & Psychology* 10(5): 639–667.
Mladenovic, B. Forthcoming. *Kuhn's Legacy: Epistemology and Metaphilosophy in Kuhn's Mature Thought.* New York: Columbia University Press.
Nickles, T. This volume. "Some Puzzles about Kuhn's Exemplars."
Read, R. 2003. "Kuhn: le Wittgenstein des sciences?" *Archives de Philosophie* 463–480.
Read, R. 2012. *Wittgenstein among the Sciences.* Farnham: Ashgate.

Sharrock, W. and Read, R. 2002. *Kuhn: Philosopher of Scientific Revolution.* Oxford: Polity.

Winch, P. [1958] 1990. *The Idea of a Social Science.* 2nd ed. London: Routledge.

Winch, P. 1964. "Understanding a Primitive Society." *American Philosophical Quarterly* 1(4): 307–324.

Winch, P. 1997. "Can We Understand Ourselves?" *Philosophical Investigations* 20(3): 193–204.

Wittgenstein, L. 1953. *Philosophical Investigations.* Oxford: Blackwell.

Wittgenstein, L. 1969. *On Certainty.* Oxford: Blackwell.

Wittgenstein, L. 1993. "Remarks on Frazer's Golden Bough." In J. Klagge and A. Nordmann (eds.), *Ludwig Wittgenstein: Philosophical Occasions.* Indianapolis, IN: Hackett Publishing Company.

Part II
Key Concepts

5 Kuhn's Paradigms

Vasso Kindi

INTRODUCTION

"Paradigm" is a key term in Thomas Kuhn's account of science as it has been articulated in *The Structure of Scientific Revolutions* (1962).[1] It has helped Kuhn advance a picture of science quite different from the one prominent at the time. "Paradigm" shifted the focus of attention from the understanding of science as theory expressed in statements, to the actual practice of science which was modeled upon a paradigm. In this chapter I will first explain how "paradigm" was used in SSR, taking into account the term's reception and Kuhn's later qualifications. I will consider a particular understanding of paradigms that has drawn the attention of contemporary scholars, namely, the one which juxtaposes them to rules. By comparing Kuhn's "paradigm" to Wittgenstein's homonymous term, I will suggest that the contrast between paradigms and rules is misleading. I will argue that paradigms set rules which, when followed, build traditions and form frameworks. In that sense the polysemy of "paradigm" is not as troublesome as has often been thought.

1. PARADIGMS IN SSR

Paradigms are defined in SSR as "accepted examples of actual scientific practice—examples which include law, theory, application, and instrumentation together—[that] provide models from which spring particular coherent traditions of scientific research" (10, cf. 23). They are the locus of professional commitment regarding standards and rules (11), they are not reducible to their components (11), they are not to be tested (122), and they are not to be corrected (122). Their function is both cognitive and normative (109). They:

- prepare students for membership in the community (11),
- permit and guide an esoteric type of research (11, 44),

- pick the class of facts that are worth determining with more precision and in a larger variety of situations (25),
- set and define the problems to be solved (27–28),
- are the criteria for choosing problems (37),
- guarantee a stable solution to these problems (28),
- breed problems and solutions (105),
- are the source of the methods, problem field, and standards of solution accepted by any mature scientific community at any given time (103),
- are the prerequisite to the discovery of laws (28),
- induce anticipations regarding phenomena (56),
- account for the observations and experiments easily accessible to the practitioners of science (62),
- lead the way for an exploration of an aspect of nature (92),
- give form to scientific life (109),
- tell the scientist about the entities nature does and does not contain and about the ways these entities behave (109),
- provide a map whose details are elucidated by research and the directions for map making (109),
- define a science (34, cf. 103),
- are constitutive of scientific activity and of nature (109–110),
- are prerequisite to perception (113),
- highlight perceptual features (125),
- provide a box into which nature can be shoved (151–152),
- are adopted largely on faith (158).[2]

As the above list illustrates, paradigms are seen as means that are instrumental in scientific education, as standards shaping scientific practice, and as the explanatory tools which are used to account for what science looks like. They are seen as operating at the level of practice (detected by the analyst in what goes on in scientific activity) but also at the level of analysis (that is, devised by the analyst to account for what goes on).

After the publication of SSR, the concept of paradigm was heavily criticized, at the same time that it was entering every possible field of study.[3] Arnold Thackray said that "it's a brilliant wrong idea" (cited in Coughlin 1982). David Hollinger (1980, 197) called it "[Kuhn's] most celebrated and maligned term" and Shapere (1980) found it mysterious, vague, and ambiguous. J. Wisdom (1974, 832) said that although it is a nice idea "it is not easy to say just what it means" while J. B. Conant wrote to Kuhn that he was afraid that he [Kuhn] will be brushed aside "as the man who grabbed on the word 'paradigm' and used it as a magic verbal wand to explain everything" (letter to Kuhn, June 5, 1961; cited in Cedarbaum 1983, 173).[4] Of course, one should not omit to mention Masterman's (1970) notorious identification of twenty-one different uses of paradigm in SSR, which largely overlap with the uses listed above.

2. KUHN'S RETROSPECTIVE ACCOUNT OF
HOW HE UNDERSTOOD PARADIGMS

In his last interview, Kuhn says that the term "was a perfectly good word until [he] messed it up" (2000, 298). What was "paradigm" before he messed it up and how did he mess it up, if, in fact, he did? In the same remark he clarifies his point and says that "[p]aradigms had been traditionally models, particularly grammatical models of the right way to do things" (298). The same claim appears in SSR: "In its established usage, a paradigm is an accepted 'model' or 'pattern'" (23). But he cautions against a particular characteristic of grammatical models: in grammar, paradigms such as "*amo, amas amat*" are imitated mechanically in conjugating other Latin verbs, for example, in producing "*laudo, laudas, laudat.*" In SSR, Kuhn says that in science, unlike grammar, and like common law, "paradigm is rarely an object of replication . . . it is an object for further articulation and specification under new or more stringent conditions" (23). When Kuhn uses "paradigm" to account for the consensus of scientists, he appropriates this aspect of "model," namely, that it is a standard which is being followed rather than imitated[5] in different conditions. The problem with his employment of the term emerges, in his view, by not restricting it to the consensus about new model applications but by extending it to cover consensus about "a hell of a lot of other things [as well] that weren't models." So, he ends up using the term "paradigm" for "the whole bloody tradition" (2000, 299).[6]

It seems that Kuhn is rejecting as problematic one of the two major senses of "paradigm" as it was introduced in SSR. In the narrow sense of the term, a paradigm is a concrete achievement or model and in the wider sense, the framework or tradition built upon it.[7] Kuhn seems to be finding fault with this latter notion but he does not explain why. In his "Postscript," to disambiguate paradigm, Kuhn speaks of "exemplar" (paradigm in the narrow sense) and "disciplinary matrix" (paradigm in the wide sense).

Several issues can be raised in relation to this account of paradigms offered by Kuhn.

• Kuhn says that he does not care for paradigms as used in grammar because, in this context, any new application can fully replace the original example chosen to function as paradigm; every new application can, in principle, obliterate the previous one and stand in its place. Kuhn prefers to treat paradigms as precedents in common law. In legal practice, every new case may differ significantly from, and may add something to, the previous ones; every new application stands next to and does not erase or substitute for older cases. Given this preference for common law rather than grammar, it is puzzling why Kuhn objects to the wide sense of paradigms. He chose to use paradigms as models which are further articulated and not mechanically imitated exactly because he wanted to get a coherent tradition out of them. Drills with grammatical paradigms may teach students to conjugate verbs (and so give rise to the relevant practice), but the examples used for the instruction of students, being instances of existing grammatical rules, are superfluous and dispensable; students can, in principle, learn to conjugate by using any other example or by invoking directly the relevant explicit rules. The examples are given to make instruction easier. In the case of science, however, as in the case of common law, paradigms are not instances of rules but, rather, concrete individual cases which are irreplaceable and indispensable in the process of instruction and initiation and in building and carrying out the relevant practice. "A paradigm is what you use when a theory [with the explicit axioms and rules] isn't there," says Kuhn, accepting Margaret Masterman's characterization (Kuhn 2000, 300). So, if Kuhn endorses understanding paradigms as building practices, it is not clear why he is reluctant to accept them as traditions. The term may be seen as used synecdochically: a particular item, the paradigm as concrete model, is used to refer to the whole tradition which is built by employing it. In this sense the two are not opposed.

- Another issue with Kuhn's account of what he did right and what he did wrong as regards the concept of paradigm is that it involves an inconsistency. In the same interview Kuhn relates how he wrote SSR:

> I wrote a chapter on revolutions, slowly but not with excessive difficulties and talking about gestalt [switches]. . . . Then I tried to write a chapter on normal science. And I kept finding that I had to—since I was taking a relatively classical, received view approach to what a scientific theory was—I had to attribute all sorts of agreement about this, that, and the other thing, which would have appeared in the axiomatization either as axioms or as definitions. And I was enough of a historian to know that that agreement did not exist among the people who were [concerned]. And that was the crucial point at which the idea of the paradigm as model entered. Once that was in place, and that was quite late in the year, the book sort of wrote itself. (Kuhn 2000, 296)

In this passage Kuhn says that he wanted to write about normal science which, by definition, involves extensive agreement about a number of things. He originally thought that this agreement would have to be about elements of scientific theories (i.e., propositions, axioms, and definitions). But, he couldn't find this kind of agreement. So, he hit upon the idea of paradigms and attributed "all sorts of agreement about this, that, and the other thing" to them. He, thus, solved his problem of writing about normal science. Yet, as was quoted earlier, he also complains that he made the mistake of using "paradigm" for more than one kind of agreement (for "a hell of a lot of other things"). There is some tension here. Kuhn, on the one hand, says that paradigms helped him account for *all sorts* of agreement in normal science and, on the other, that paradigms are responsible for only one sort of agreement, namely, agreement regarding model applications. Any exten-sion of the use of paradigm to cover agreement about other things is, in his view, illegitimate and results in messing up the use of the term.

• A third problem with Kuhn's account of what he did with paradigms appears when one considers that Kuhn, as noted earlier, proceeded to dif-ferentiate between "exemplar" (paradigm in the narrow sense) and "dis-ciplinary matrix" (paradigm in the wide sense) in his "Postscript." If he objects to the wide use of "paradigm," that is, paradigm as tradition, why did he care to give it a different name in the "Postscript"? Why didn't he reject the concept of paradigm-as-tradition altogether? It seems that his only concern was to dispel the criticism of polysemy.

The above problems show that Kuhn's retrospective account of what went wrong with "paradigm" is not very satisfactory. Kuhn maintained that he should not have allowed the term to cover both exemplar and framework. The first remedy he offered already in the "Postscript" was to disambiguate the term, distinguishing between exemplar and disciplinary matrix. But he came to realize that this was not really what was causing trouble and dropped the use of "disciplinary matrix" altogether in his later writings (cf. Hoyningen-Huene 1993, 132).[8] He didn't have much use for "exemplar" either, preferring, instead, terms such as "lexicon," "taxonomy," "lan-guage," and "system," which, however, are more easily associated with the wider sense of "paradigm," a sense that he originally said he was interested to remove and which he took to belong to the level of language rather than that of practice. Under the pressure of criticism, which focused on issues of meaning variance and incommensurability of concepts, Kuhn increasingly concentrated on understanding the concept of paradigm linguistically, leav-ing behind the more practical dimensions which he originally introduced. Lexicons were taken to be wholes, consisting of concepts and terms form-ing sentences. In reconstructing his account, Kuhn proceeded to disown the concept of paradigm in the wide sense of tradition and acknowledged only the more dignified conception of paradigm as exemplar or model. But that move, which privileged exemplar over tradition, did not really make a dif-ference in his overall scheme and, as we saw above, did not really smooth out tensions and inconsistencies. What is more, the concept of exemplar was not anymore very relevant to what Kuhn did and, consequently, it even-tually fell out of his focus.[9] Other scholars, however, maintained that para-digm as exemplar is "Kuhn's most important concept" (Forrester 2007, 783; cf. Crane 1980, 33; Nickles 1998, 2003).[10]

In what follows, I will argue, *pace* later Kuhn, that the two notions (paradigm as exemplar and paradigm as tradition[11]) are very much con-nected and that he did not really mess up the term, at least in the original formulation in SSR. I will maintain that the use of exemplars sets rules which, when followed, establish a practice which eventually forms a tradi-tion and a framework. In that sense, exemplars should not be contrasted to rules, as several scholars have lately claimed. Wittgenstein's philosophy, which influenced Kuhn's work, will help me show these connections. I will begin by presenting how Kuhn came to use the term "paradigm" and then proceed to compare Kuhn's understanding of "paradigm" to Wittgenstein's use of the same term.

3. HOW DID KUHN COME TO USE "PARADIGM"?

We, first, need to distinguish between appropriating the term and getting the notion. The term, in its standard sense of prototype or paradigmatic sample, was widely used in every field of inquiry and appears already in Kuhn's *The Copernican Revolution* (1957). He says there that the cases from the history of science discussed at the General Education courses at Harvard College functioned "as paradigms [of science in history] rather than being intrinsically useful bits of information" (ix). Also, discussing telescopic observations of dark spots on the surface of the sun by Galileo, Kuhn says in the same book that "the motion of the spots across the sun's disk indicated that the sun rotated continually on its axis and thus provided a visible paradigm for the axial rotation of the earth" (221–222). In the same period, the term appears, in this standard sense, in most of the books Kuhn mentions in SSR: in Hanson's *Patterns of Discovery*, in Whorf's *Language, Thought and Reality*, in Polanyi's *Personal Knowledge*, and in Fleck's *Entstehung und Entwicklung einer wissenschaftlichen Tatsache*. Given all this, there is nothing special to explain regarding the employment of the term "paradigm" by Kuhn in its standard sense. What about the nonstandard use of the term in SSR?[12]

In his Preface to *The Essential Tension* (1977), Kuhn explains that he first used the term "paradigm" in the sense he eventually adopted in SSR, in his essay "The Essential Tension" (Kuhn 1977a), which was read at a conference at the University of Utah in 1959. The topic of the conference was creativity and scientific talent, and the participants, mostly psychologists, emphasized the importance of imagination, freedom, and open-mindedness. Against this view, which expresses the typical understanding of scientists, that is, as the intrepid, free spirits who conquer the frontiers of knowledge, Kuhn advanced the highly controversial claim that scientific creativity depends on convergent thinking which is achieved by a dogmatic initiation and a rigid education centered on repetitive exercises with concrete problem solutions, the paradigms.[13] It seems that Kuhn formulated his concept of paradigm (adopting also the term), drawing on his experience as a physicist but also as a historian. He had been subject himself to the education he described and he knew from his research in the history of science that "work within a well-defined and deeply ingrained tradition seems more productive of tradition-shattering novelties than work in which no similarly convergent standards are involved" (234).

K. Brad Wray, in his paper about the discovery of the idea of paradigm (Wray 2011), claims that Kuhn offers two different stories regarding the origin of his concept: one that focuses on the differences between natural and social sciences (natural sciences depend on paradigms while social sciences do not[14]) and another that focuses on the consensus necessary for effective research. Wray says that since Kuhn gives these two different stories, Kuhn himself may be mistaken about his own discovery (2). I don't see how this follows (one of the stories may be correct) but, most crucially, I don't see why the stories are different. I take it that the two accounts are complementary. Kuhn was struck by the fact that there is consensus in the practice of the natural sciences, something which is not the case with the social sciences, but he couldn't find evidence of agreement regarding explicit rules or definitions. At some point, he realized that the key was the type of education and he hit upon the word and concept of paradigm. That proved to be the "missing element" necessary to finish the book (Kuhn 1977, xix).

Daniel Cedarbaum, in "Paradigms" (1983), advances the view that Kuhn took "paradigm" from Wittgenstein, who was influenced, in turn, by Georg Lichtenberg. Cedarbaum states explicitly that Kuhn, whom he had interviewed, "does not remember taking the term "paradigm" from Wittgenstein" but insists that "Wittgenstein's treatment of naming in the *Philosophical Investigations* may have had a crucial impact on [Kuhn's] formulation of the paradigm concept in the spring of 1959."[15] Cedarbaum cites as evidence Kuhn's references to Wittgenstein's account of naming in SSR and mentions Stanley Cavell as quite possibly a major influence on Kuhn in that respect.[16] In his last interview, Kuhn denies categorically that he knew of Lichtenberg's or Wittgenstein's use of the paradigm concept: "I certainly was not aware of either of them. Lichtenberg was called to my attention [presumably by Cedarbaum], and I am a little surprised that I haven't had my nose dragged through Wittgenstein's use of it."(2000, 299). Cavell, however, recalls vividly in his autobiography (2010, 354) that Kuhn had told him at Berkeley that he [Kuhn] knew that Wittgenstein uses the idea of paradigm.[17]

It is difficult to settle the historical question of whether Kuhn did in fact get the concept and the term "paradigm" from Wittgenstein. He certainly knew of the relevant material but we cannot tell whether he got the actual term and notion from Wittgenstein. The reason, I think, Wittgenstein's use of the term "paradigm" was not registered by Kuhn is that Wittgenstein's concept, unlike, for instance, his concepts of language game or form of life, was very little discussed in the literature. There were brief and sporadic comments regarding similarities and differences between Kuhn's and Wittgenstein's term, mostly from within philosophy of science,[18] but there wasn't really any sustained discussion of Wittgenstein's "paradigm" in the secondary literature on his philosophy.[19]

One place where Wittgenstein's "paradigm" was used and, in fact, in relation to science, was Stephen Toulmin's book *Foresight and Understanding*, which was published in 1961, just before SSR's publication. The term "paradigm" appears in its pages repeatedly. Toulmin, clearly, borrows the term from Wittgenstein and takes it to mean "object of comparison," again a Wittgensteinian term. Actually, Toulmin uses Wittgensteinian ideas throughout the book, and applies them to science. For instance, he argues that science cannot have a single aim and purpose that can be captured by a definition, echoing Wittgenstein's opposition to essentialism. Particularly telling is the comparison he makes between science and sport, which is very much reminiscent of Wittgenstein's discussion of games, as is Toulmin's claim that explanations reach rock bottom (1961, 42), a phrase very similar to Wittgenstein's idea that justifications reach bedrock where spades are turned (PI, 217).[20]

Apart from "objects of comparison" (Toulmin 1961, 52), paradigms for Toulmin are "ideals of natural order" (38), models and ideals as well as principles of regularity (42–43), fundamental patterns of expectation (47, 56), explanatory conceptions (52), "standards of rationality and intelligibility" (56), standard cases (57), intellectual patterns which define the range of things we can accept (81), preconceived notions (100). They stand to reason (42), are self-explanatory (42), set the regular order of things and what departs from it and needs explanation (54, 79), are not true or false (57), "take us further (or less far)" and are more or less "fruitful" (57). They change and develop and are broadly empirical, but one cannot confront them directly with the results of observation and experiment (100). Our commitments to them blind us to other possibilities (101). Those who accept different ideals have no common theoretical terms (57) while cross-type comparisons are not fair (62).

Kuhn does not compare his use of "paradigm" to Toulmin's in any of his writings. He says that he deliberately did not read *Foresight and Understanding* while he was trying to write SSR,[21] but he also admits, not referring to "paradigm" in particular, that he understands "why Toulmin might have been sore at me for stealing his ideas" (2000, 297). Toulmin, however, does not complain of any such thing.[22] In fact, he says that Wittgenstein's "theory of paradigms," which he also advocated, is very different from Kuhn's since Wittgenstein's, and his own, theory does not imply discontinuous change (Toulmin 1972, 106–107).[23]

In what follows I will not try to establish a detailed line of influence from Wittgenstein to Kuhn,[24] but, rather, I will argue that Kuhn's concept of paradigm has indeed much in common with and draws upon Wittgensteinian considerations regarding rule following, which also involve a discussion of paradigms and examples. More than simply establishing the affinities, I will argue that seeing Kuhn's paradigm from a Wittgensteinian perspective lends support to the original conception of paradigm by Kuhn, which combines both the narrow (exemplar) and the wide (framework) sense of the term. Accordingly, I will maintain that the contrast between exemplar and rule, which appears in Kuhn's work and features prominently in recent secondary literature on Kuhn, is misleading and rests on a very particular understanding of the concept of rule.[25]

4. THE PRIORITY OF PARADIGMS IN SSR—THE CONNECTION WITH WITTGENSTEIN

In SSR Kuhn devotes a chapter to what he calls the priority of paradigms [over rules] in which he compares his discussion of paradigms to Wittgenstein's discussion of applying terms "unequivocally and without provoking argument" (45). His main concern in this chapter is to show that scientific activity, and the cohesion of scientific tradition, does not depend on the operation of rules but can rather be attributed to (and is, in fact, dependent upon) the operation of paradigms regardless of the existence of rules. He identifies paradigms with the standard illustrations of theories in their various applications which recur in textbooks, lectures, and laboratory exercises, while he understands rules to be isolable elements abstracted from paradigms and articulated retrospectively by scientists, that is, after they have been initiated into the profession and usually only if they are asked to rationalize or interpret what they do.

Kuhn gives four arguments in favor of the view that paradigms not only could function without the presence of rules, but do, in fact, function that way. The first argument notes the difficulty of discovering rules governing scientific practice. Just as in the case of games that Wittgenstein discusses, it is difficult to find their essential characteristics—what all games have in common; in a similar manner, Kuhn says, it is difficult to find the rules which can account for the correct practice of science.[26] Kuhn wants to say that just as we do not need the common, essential, characteristics of things to use the words referring to them correctly, in the same way we do not need rules to practice science properly. The second argument says that scientists do not actually come across rules in their scientific education and so, he implies, they do not need them. Scientists, Kuhn says, do not learn theories in the abstract, that is, by following abstract rules; rather, they study particular examples of theory applications (the paradigms) and do finger exercises with them. So, contrary to what is commonly believed about examples and paradigms in general, that is, that they are only needed to either document general propositions or illustrate rules, Kuhn maintains that paradigms are anything but dispensable. "They are not there merely as embroidery or even as documentation" (46–47). They have a vital role to play and, according to Kuhn, it is rules that are superfluous. The third argument focuses on the fact that rules become important only during periods of crisis. When scientists proceed securely in their practice following the paradigms of their discipline, they have no need for rules. This means that paradigms are prior to rules in the sense that paradigms need to be presupposed in order even for rules to become important in the rare occasion of a crisis. The fourth and last argument aims to show that paradigms, unlike rules, are necessary to account for the diversity of scientific fields. "Explicit rules, when they exist, are usually common to a very broad scientific group, but paradigms need not be" (49). In the process of specialization scientists practice with and follow different paradigms even if they may presuppose common rules.[27]

In the course of this discussion, Kuhn invokes Wittgenstein to answer two questions: (1) what restricts scientists to a particular normal-scientific tradition in the absence of rules and (2) what can the phrase "direct inspection of a paradigm" mean? (44). In the case of Wittgenstein, as presented by Kuhn, the use of words is not restricted by some set of characteristics common to all members of the class referred to by the particular terms, but rather by a network of crisscross similarities which form natural families.[28] In the same way, the research problems addressed by the scientists in a certain tradition need not have anything in common (for instance, falling under some general rule). They may, instead, relate to each other by crisscross similarities and by having been modeled upon the accepted paradigms. By appealing to Wittgenstein, Kuhn solved his problem of accounting for the cohesion of a tradition in view of the difficulty he had in discovering any rules commonly adopted by scientists. So, the answer to the first question is "paradigms"; that's what restricts scientists to a particular course of practice. The answer to the second question is "identifying similarities and modeling a problem to a paradigm"; that's what "direct inspection of a paradigm" means.

So, Kuhn's preference for paradigms over rules in SSR was instigated by the fact that he could not discover rules which scientists would cite and follow to produce scientific knowledge. What kind of rules could these be? Methodological rules, for instance, how to experiment, how to make inferences, how to test laws, how to collect data, how to improve accuracy, how to expand a series, etc.; explicit definitions of scientific terms which would guide their implementation, but also elements of theories, such as laws or generalizations, which would cover all particular instances of application. For instance, a law of the form "all As are Bs" can be interpreted as a rule which dictates that all As ought to be also Bs. Having found no consensus upon these among scientists, Kuhn turned to paradigms.[29] It's not that scientists could cite paradigms instead of rules. Kuhn discovered paradigms looking into textbooks and scientific education. Instead of concentrating on what the philosophers described and prescribed as the scientific method, he concentrated on scientific practice and found particular problem solutions repeatedly used in textbooks and in training. These exemplary problems and solutions, which embody all the elements assigned to the disciplinary matrix later,[30] that is, both methodological and theoretical items, he called "paradigms." Kuhn did not claim, as Feyerabend did (1979, 18), that scientists do not follow rules because they are "unscrupulous opportunists," applying whatever methodology happens to suit the occasion.[31] Rather,

Kuhn turned his attention from an intellectualist, epistemological under-standing of science, in terms of rigid formal rules, to a practical one which highlights the concrete idiosyncrasies of a practice. Feyerabend, a liberal pluralist, rejected the uniformity of method enjoined by the epistemologists and allowed more options; Kuhn gave up this picture of choosing from epistemological alternatives altogether and attributed agreement in practice to the type of training.

Subsequently in the literature, the privileging of exemplars over rules was taken out of the practical context of scientific education into a theo-retical cognitive one and was understood as promoting a certain way of learning and reasoning, for instance, case-based over rule-based. Alexan-der Bird credits Kuhn with the thought that "not all human psychology can be explained in terms of rules and expert systems, and that in particular certain kinds of learning through concrete examples need not be seen in such terms (Bird 2000, 74). Thomas Nickles contended that "for Kuhn scientific methodology (insofar as that enterprise can be defended at all) is a case-based rather than a rule-based system" (Nickles 2000, 244),[32] although he admits that Kuhn says little about the representation of exem-plars in human cognition (ibid., 249).

Kuhn did not say much about the role of exemplars in human cogni-tion because this would mean that he would be engaged in doing psychol-ogy rather than philosophy. He experimented in this direction in the early period after SSR, but he eventually returned to the philosophical issues that mostly concerned him, namely, the problems of incommensurability and meaning change. Also, Kuhn's description of scientific methodology as case-based rather than rule-based is ambiguous. If the reference is to case-based, rather than rule-based, *reasoning*, then we credit Kuhn with an interest that he did not really have, namely, an interest in modeling, for instance, human reasoning, much like current work in the fields of cogni-tive science and artificial intelligence. If, on the other hand, a comparison is made between Kuhn's description of scientific methodology as case-based and the methodology followed in case-based disciplines, such as law, medi-cine, or psychoanalysis, then certain differences become salient. For one, the particular cases in the case-based disciplines are themselves the focus of attention; they are not treated as instances of some general rule. In both law and medicine, for instance, one is interested in describing and solving problems that pertain to the particular cases under consideration. Even precedents, which are appealed to in order to provide guidance for future decisions, are studied in detail so that, in later cases, specific similarities and differences between them are identified and assessed. In scientific meth-odology, on the other hand, as described by Kuhn, paradigmatic cases, the paradigms, are not studied for what they are in themselves, but as teaching tools for future action. Wittgenstein put the difference as follows: "Teach-ing which is not meant to apply to anything but the examples given is dif-ferent from that which '*points beyond*' them." (PI 208).[33] In case-based disciplines, the cases used as precedents (exemplars) need to be studied in detail in order to be able to tell whether a new case is similar to the precedent and can be ruled or treated similarly. In science training as described by Kuhn, however, exemplars are used to point beyond them. That is, students who learn projectile motion, for instance, do not need to concentrate on the particulars of the examples used in teaching it; for example, they do not need to remember the specific values of the variables in the original setting of the problem situation. They are asked to practice with exercises which are mapped onto, and are, therefore, *already* taken to be similar to, the exemplary one used in teaching. By solving these, students develop the skill to map future problems onto the original one and treat them accordingly. In case-based disciplines, we don't take it for granted but are trying to discover (or establish) whether the later case is similar to precedent. We know the specific characteristics of precedent and then check whether the case under consideration is similar to the original one. In science, always as described by Kuhn, the similarity between precedent and later case is not discovered post hoc but already given or imposed by training (cf. Lipton 2005, 1264). What one acquires by being exposed to Kuhnian paradigms is the skill to transcend the particular exemplary case and move on to future cases guided by it. This is an idea Kuhn found in Wittgenstein.

5. WITTGENSTEIN'S PARADIGMS AND RULES

Wittgenstein uses the term "Paradigma" in nearly all his published writings in the sense of sample (BB, 128; WL, 143), prototype (RFM III, 7), model (RFM III, 31, 41), standard (RFM VI, 22), object of comparison (BB, 166; PI, §50; PR, 57), rule (PG, 419; RFM III 28; LFM, 55).[34] A paradigm can be something special, like the standard meter kept in Paris, or just any example of a specific practice (for instance, an example of language usage, of mathematical proof, or of behavior) that is chosen as a means of instruction in order to be followed on future occasions. It is something particular, a concrete case which has, however, a general import. It lays down what is correct to do by being itself the measure of correctness. The idea is that a sample of something is used to show how one is to go on in accordance with it. Instead of having a general formula, or an injunction expressed in words, which tells us what to do, we have a concrete instantiation of, say, some color, or of an arithmetical calculation which we learn to follow. Several problems immediately arise: (1) What is the relation between the samples used in instruction and the rules they set? (2) How do the rules compel us in a certain direction? (3) How are we to know which of the properties not only possessed by the sample used for instruction, but also exemplified by it,[35] are to be concentrated on and followed? When a child, for instance, is shown a swatch of blue cloth and told "blue," how is she to know whether the term refers to the color, the texture, the shape, or any other quality the swatch exemplifies and use the term accordingly in the future? (4) How is one guided to move from the original sample, or paradigm, to all the other cases that are supposed to be treated as similar to it? How does one read off the instruction and how is generality achieved? (5) Is the similarity between paradigm and future cases something waiting to be recognized or is the similarity imposed and established in practice?

All these issues are taken up by Wittgenstein in his discussion of rule following and by scholars in the vast secondary literature on the topic. I will not review the discussion and controversies here; what I would like to do is to sketch briefly how I see Wittgenstein's understanding of rule following in order to illuminate Kuhn's understanding of paradigms. I would like to appropriate Wittgenstein's work on the issue to make three points: (1) discussion of paradigms is not to be seen as dealing with an empirical account of how human beings follow rules nor with modeling human reasoning, but rather with issues that pertain to how rules relate to practice; (2) paradigms should not be contrasted to rules; and (3) it is not a mistake to understand paradigms both in a narrow and in a wide sense.

Wittgenstein understands prototypes or paradigms as models which articulate a particular way of conceiving things: "the prototype must be presented for what it is; as characterizing the whole examination and determining its form. In this way it stands at the head & is generally valid by virtue of determining the form of examination, not by virtue of a claim that everything which is true only of it holds for all the objects to which the examination is applied." (Manuscript 111, 119–120; referred to in Kuusela 2008, 124). That is, paradigms set the stage, open up a space in which things are supposed to be done in the way exemplified by the paradigm. This does not mean that one must copy or reproduce exactly what the paradigm says or looks like; rather one is supposed to move forward by assimilating further cases to the exemplary ones used in instruction. As Wittgenstein put it, paradigms are "centres of variation" (Manuscript 152, 16–17; referred to in Kuusela 2008, 173). Once the example of a prototype is followed, a rule is (or rules are) set and a particular practice emerges. For instance, being exposed to samples of color one learns what color is and how to use the words for color. By following the rules derived from the samples (on what occasions one uses the relevant words), one learns to participate in the practice of using color words (referring to color, identifying color, distinguishing colors, etc.), which involves also extending the practice. Thus, the original paradigm (the particular exemplar), is responsible for the development of a whole practice. It's like a point which expands into a whole open-ended area.

Wittgenstein's discussion of rules and paradigms is not empirical. He did not seek to uncover facts regarding the way human beings learn and follow rules. Nor did he try to model human reasoning. His effort aimed at sketching the grammar of the relevant notions. Kuhn should be seen as doing something similar. He was not involved in a psychological or sociological investigation but rather in a logical one informed by history, which offered an account of the conditions that make science possible (cf. Kindi 2005). If we see Kuhn's paradigms under these Wittgensteinian lights, then we can also see how it is possible to understand paradigms both in the narrow (exemplar) and the wide (practice-tradition) sense. As regards the relation of paradigms to rules we saw that Kuhn, and later commentators, contrasted the two notions. However, Wittgenstein's analysis may help us clarify what Kuhn meant and allow for a more complex understanding.

Traditionally, it was supposed that one learns what a word or a formula means by being given a definition which functions as a rule. For instance, "horse" means a large four-legged animal with a mane. Given this definition, one knows how to proceed, that is, how to use the word "horse" and how to identify horses. Understanding the definition, or any kind of instruction, meant understanding the words comprising it. But we cannot go on to explain words by words for ever. As Schlick put it,

> In order to arrive at the meaning of a sentence or a proposition we must go beyond propositions. For we cannot hope to explain the meaning of a proposition merely by presenting another proposition. . . . I could always go on asking 'But what does this proposition mean?' You see there would never be any end to this kind of enquiry, the meaning would never be clarified if there were no other way of defining it than by a series of propositions. . . . The discovery of the meaning of any proposition must ultimately be achieved by some act, some immediate procedure, for instance, as the showing of yellow; it cannot be given in a proposition. (cited in Hanfling 1981, 19)

Connecting words to samples was considered to be one way of linking language to the world. But given the problems of exemplification noted by Goodman (which of the properties possessed by the sample are exemplified?), samples cannot unproblematically play the role of anchoring language. Just pointing to a sample does not tell us what features to concentrate on and what to do to follow it. Wittgenstein took samples to be already symbols, that is, already connected to the world through a particular linguistic, or other, practice. The way a sample is copied and followed cannot just be dictated by the sample *qua* physical object but has to be learned in training and picked up in practice.

Samples, or exemplars, set rules but also, through their use in a particular practice, anchor rules to reality and restrict their application. Rules cannot determine by themselves how they ought to be followed. We would be involved in an infinite regress if we were to add rules to rules to learn how to employ them. We need examples of application to break the impasse and learn how to go on. If practical training with examples was missing, then any course of action could be shown to be consistent with the expression of the rule, which means that "there would be neither accord nor conflict

here" (PI, §198–202). We would be trapped in an infinite regress of interpretations, substituting one expression of the rule for another. What makes
the following of a rule possible, in a specific and concrete manner, is training with actual cases.

Not only rules, but also examples are needed for establishing a practice. Our rules leave loopholes open, and the practice has to speak for
itself (OC, 139).[36] So, rules need examples/exemplars in order to be properly followed; practice with exemplars sets rules which, as they are followed, form traditions.[37]

6. CONCLUSION

The complex understanding of the relation between exemplar and rule
offered by Wittgenstein fits very well, I think, Kuhn's use of the relevant
notions. The contrast between exemplar and rule which we find in Kuhn's
work should not be seen as a general opposition but, rather, as a contrast
between following particular examples of application and following abstract
expressions of general methodological rules dictated by a particular philosophical understanding of science. Kuhn maintained that scientists do not
follow the abstract rules that epistemologists used to prescribe. They are
rigorously trained to emulate concrete examples of scientific achievement.
This practice gives rise to a rule-governed behavior which eventually builds
a tradition. The rules followed are not general prescriptions of the form
"confirm your hypotheses" but, rather, patterns of behavior, closely modeled on exemplars, which have been mastered practically in education and
carried over in research. The mutual dependence of exemplar and rule and
their common connection to tradition show that the notorious polysemy of
"paradigm" is not to be eradicated as pernicious. The different senses of the
term (for instance, exemplar and framework or tradition) reflect the different functions performed by paradigms.

NOTES

1. Abbreviated as SSR. All references to this book are given directly by page
   numbers.
2. In the above list I kept (with occasional editing) Kuhn's phraseology.
3. See, for instance, the indicative study of Cohen (1999).
4. Cf.: "Unfortunately that word [paradigm] has been used in so many senses,
   not least in Kuhn's (1970) classic discussion itself, that its meaning has been
   almost blurred out of existence" (Lipton 2005, 1264); "Kuhn's portmanteau
   notion of the 'paradigm'" (Daston 2010, 217).
5. Cf. the distinction between imitation and following made by Kant in §32
   and §49 of his Third Critique (Kant 2000). Martin Gammon (1997) extensively discusses the differences Kant notes between imitation (*Nachahmung*),
   following (*Nachfolge*), mechanical replication (*Nachmachung*), and aping

(*Nachäffung*). According to Gammon (1997, 586), Kant also distinguishes between two senses of the exemplary: as an archetype (*Urbild*) for emulation and as a pattern (*Muster*) for imitation. No copy is adequate to the archetype; it is the original measure of things. Kant sees the work of genius in fine art "as an archetype (*Urbild*) for the emulation (*Nachfolge*) of future geniuses, as a pattern (*Muster*) for the imitation (*Nachahmung*) of future artists, as a model (*Modell*) for the replication (*Nachmachung*) by schools and as an expression of peculiarity (*Eigenthumlichkeit*) which may serve for the aping (*Nachäffung*) of counterfeits, plagiarists, and 'tyros'" (ibid., 588). The work of a genius sets a new rule not for future geniuses but for aesthetic instruction in schools for future artists. In science, Kant says, we can learn the work of great inventors, who differ from an imitator and apprentice only in degree, "but we cannot learn to write in truly poetic vein" (ibid., 590).

6. Kuhn does not specify where this sleight of hand took place, whether in SSR or later. Judging from what he says in the Preface of *The Essential Tension* (1977, xviii), it seems that the expansion of the concept took place even before the publication of SSR. "That concept [paradigm] had come to me only a few months before the paper ["The Essential Tension"] was read [in June 1959], and by the time I employed it again in 1961 and 1962 [in the first draft of SSR and in "The Function of Dogma in Scientific Research"] its content had expanded to global proportions, disguising my original intent." "[P]aradigms took on a life of their own" (ibid., xix).

7. Toulmin (1963, 384) understands differently the distinction between a narrower and a wider sense of paradigm. Paradigm in the narrow sense refers to a particular set of basic concepts while paradigm in the wider sense refers to a whole masterpiece of science.

8. In a letter to Donald Gillies (2 April 1990, Kuhn Archive, Box 20) Kuhn expresses his preference for the term "research programme" as substitute for the wider sense of paradigms: "I've often wished that I'd thought of the term 'research programme' for the more popular of the two different senses in which I used 'paradigm.'"

9. Wray claimed that, although theory change became taxonomic change in Kuhn's later writings, exemplars did not become obsolete or irrelevant to our understanding of science since they continued to be important to how theories are learned (Wray 2011, 392). Wray, however, neglects to mention that Kuhn was not as much interested in scientific education as he had been when he wrote SSR.

10. It's not clear, however, whether their claim relates to the significance of exemplar in general or as regards Kuhn's philosophy only.

11. Disciplinary matrix and tradition do not exactly mean the same thing. I take disciplinary matrix to be a logical notion and tradition to be an epistemological and sociological one. Disciplinary matrix is introduced to provide the context in which specific elements, such as generalizations, models, values, and exemplars, are orderly arranged. The concept of tradition refers to how knowledge is transmitted and put to use in the practice of a scientific community.

12. On June 29, 1961, in his letter to J. B. Conant, where he responds to the latter's criticism of paradigm, Kuhn says that his usage, "though certainly not normal, . . . does not strain the dictionary definition as much as [Conant] impl[ies]" (Kuhn Archive, Box 25; cf. Kuhn 1977, xix).

13. Kuhn presents the same ideas in "The Function of Dogma in Scientific Research" (1963; 1963a), read at Oxford in July of 1961.

14. Read and Sharrock (this volume) claim that the key to understanding Kuhn's work is to reflect on the contrast between the social and the natural sciences.

15. At that time, Kuhn found himself at an impasse trying to account for the consensus in the period between scientific revolutions.
16. Cf. Kindi (2010) for a slightly different account of the relation between Kuhn and Cavell.
17. It is worth quoting the passage in full: " . . . Kuhn, perhaps after a department meeting, accompanied me home for a drink, and, talking past midnight Tom was becoming agitated in a way I had not seen. He suddenly lurched forward in his chair with a somewhat tortured look that I had begun to be familiar with. 'I know Wittgenstein uses the idea of "paradigm". But I do not see its implications in his work. How do I answer the objection that this destroys the truth of science? I deplore the idea'" (Cavell 2010, 354–355).
18. Toulmin (1973, 284n12) and in a certain sense Stegmüller (1976, 170ff.).
19. The only exceptions that I am aware of are Toulmin (1961), where the term is used rather than analyzed, Luckhardt (1978), Kindi (1995), and Kindi and Zika (2005).
20. None of the reviews of *Foresight and Understanding*, however, mentions the connection to Wittgenstein (see Achinstein 1963; Cooper 1963; Kyburg 1963). Toulmin acknowledges his "special debt to the late Professor Ludwig Wittgenstein" in his Preface to his earlier book *The Philosophy of Science: An Introduction* (1953).
21. In the same place he says that he hadn't read Polanyi's book either, but he cites the book in SSR, pointing to two particular chapters. In general, I remember also from a personal communication that he avoided reading books relevant to what he did from fear, as he said, that he wouldn't be able (he wouldn't have the time) to formulate and express his own thoughts.
22. For instance, Toulmin does not connect his and Kuhn's use of "paradigm" in his commentary on Kuhn's work at Bedford College in 1965 (Toulmin 1970), while in *Human Understanding* (1972, 100–101) he explicitly compares Kuhn's paradigms to Collingwood's absolute presuppositions.
23. Cf. "The use of this term [paradigm] by Wittgenstein himself . . . differs significantly from that made familiar recently by T. S. Kuhn in his much discussed book, *The Structure of Scientific Revolutions*" (Janik and Toulmin 1973, 284n12).
24. Wray (2011) offers a history of Kuhn's discovery of paradigms.
25. Nickles (this volume) also discusses Kuhn's understanding of exemplars and their role in establishing the practice of normal science. Nickles' account is less exegetical in comparison to mine and aims at showing that normal science is not a static phase, but that it evolves, has a structure, and merges with extraordinary science. According to Nickles, exemplars themselves also evolve.
26. Here, it should be noted that mention of difficulty does not imply that these common characteristics or the rules exist, but are only difficult to find.
27. Here Kuhn gives the example of the physical scientists who all learn the laws of quantum mechanics by being given, however, different paradigm applications in their various special fields. It seems that in this example Kuhn identifies rules and laws as being both abstract.
28. It is noteworthy that Kuhn expresses here for the first time a concern that will come to obsess him later in his work. He worries about how the world features in and channels our practices. "Wittgenstein . . . says almost nothing about the sort of world necessary to support the naming procedure he outlines" (45n2). Although I agree with Read and Sharrock (this volume) that Kuhn is not in general interested in metaphysical issues, Hanne Andersen (this volume) is right in suggesting that, in the case of family resemblance concepts, Kuhn moves beyond Wittgenstein when he formulates a particular

condition on the world in order to account for our success in identifying the objects and activities corresponding to the family resemblance concepts.

29. Peter Lipton described the situation as follows: Kuhn, having turned from physics to the history of science, was struck by the consensus he observed among scientists during periods of normal science. "It is as if all the scientists in the group had the same secret rulebook for doing good science in their specialty. The nonexistence of the rulebook gave Kuhn his question: how does one explain the rule-like behaviour of a scientific community in the absence of rules? Kuhn's answer: by exemplars" (2005, 1264).

30. As noted earlier, Kuhn in his "Postscript" (182–187) distinguishes between exemplar and disciplinary matrix but includes exemplars as elements of the matrix. The other elements are symbolic generalizations, models, and values.

31. Feyerabend (1979, 18) quotes Einstein, who said that the scientist "must appear to the systematic epistemologist as a type of unscrupulous opportunist."

32. To be sure, Nickles also considers Kuhn a schema theorist, that is, as adopting the view that past experience shapes future experience via schemata, that is, large organizing structures, such as the large paradigms or disciplinary matrices, which are created by this past experience (2000, 246, 247).

33. Cf. "[In] an ostensive definition I do not state anything about the paradigm (sample), but only use it to make statements, that it belongs to the symbolism and is not one of the objects to which I apply the symbolism" (BT, 408).

34. The particular references to these different senses of "paradigm" in Wittgenstein's philosophy are merely indicative since similar uses can be found in many other places in his published work. The term "exemplar" is rarely employed. In one place, PI §272, the German term *Exemplar* is better translated as specimen (see the new translation by Hacker and Schulte), while in Z 444 the English 'exemplar' is used by Anscombe to translate *Urbild*.

35. Cf. Nelson Goodman's discussion of exemplification (1976, 52–57).

36. Kant had already seen this. In the *Critique of Pure Reason* (A133–134/B172–173), he claims that we cannot add rules to rules to learn how to employ them. That would involve us in an infinite regress. We need adequate training with particular examples to learn how to go on.

37. One might say that a particular example is as mute as the expression of a rule as regards the way it is to be followed. But an example (or a sample) which is made part of such a process of instruction is not a mere object. It is an object (be it a spoken or a written word, a sample, an experimental apparatus, an expression of a scientific law) put to a specific use which concretely illustrates the way it is to be further applied. Students pick it up as teachers muster a number of techniques to beat students into line (PI 208). The course of development in subsequent applications of the examples learned may not be smooth and may branch out in different directions which may involve the introduction of new examples of use.

REFERENCES

Achinstein, Peter. 1963. Review of *Foresight and Understanding*, by Stephen Toulmin. *Isis* 54: 408–410.

Andersen, Hanne. This volume. "Scientific Concepts and Conceptual Change."

Bird, Alexander. 2000. *Thomas Kuhn*. Chesham: Acumen

Cavell, S. 2010. *Little Did I Know: Excerpts from Memory*. Stanford, CA: Stanford University Press.

Cedarbaum, D. G. 1983. "Paradigms." *Studies in the History and Philosophy of Science* 14(3): 173–213.

Cohen, Jon. 1999. "The March of Paradigms." *Science*, New Series, 283(5410): 1998–1999.

Cooper, Neil. 1963. Review of *Foresight and Understanding*, by Stephen Toulmin. *The Philosophical Quarterly* 13(51): 180–181.

Coughlin, Ellen K. 1982. "Thomas Kuhn's Ideas about Science. 20 Years after the Revolution." *The Chronicle of Higher Education*, Sept. 22, 1982: 21–23.

Crane, Diana. 1980. "An Exploratory Study of Kuhnian Paradigms in Theoretical High Energy Physics." *Social Studies of Science* 10(1): 23–54.

Daston, Lorraine. 2010. "Human Nature Is a Garden." *Interdisciplinary Science Reviews* 35(3–4): 215–230.

Feyerabend, Paul. 1979. *Against Method*. London: Verso.

Forrester, John. 2007. "On Kuhn's Case: Psychoanalysis and the Paradigm." *Critical Inquiry* 33: 782–819.

Gammon, Martin. 1997. "Exemplary Originality." *Journal of the History of Philosophy* 35(4): 563–592.

Goodman, Nelson. 1976. *Languages of Art*. Indianapolis, IN: Hackett.

Hanfling, Oswald. 1981. *Logical Positivism*. Oxford: Blackwell.

Hanson, N. R. 1958. *Patterns of Discovery*. Cambridge: Cambridge University Press.

Hollinger, David. 1980. "T. S. Kuhn's Theory of Science and Its Implications for History." In G. Gutting (ed.), *Paradigms and Revolutions*. Notre Dame: University of Notre Dame Press, 195–222.

Janik, Alan, and Stephen Toulmin. 1973. *Wittgenstein's Vienna*. New York: Touchstone Book.

Kant, I. 1990. *The Critique of Pure Reason*. Trans. Norman Kemp Smith. New York: Macmillan.

Kant, I. 2000. *Critique of the Power of Judgment*. Ed. Paul Gauyer. Trans. Paul Guyer and Eric Matthews. Cambridge: Cambridge University Press.

Kindi, V. 1995. *Kuhn and Wittgenstein: Philosophical Investigation of The Structure of Scientific Revolutions*. Athens, Greece: Smili Editions (in Greek).

Kindi, V., and F. Zika. 2005. "Paradigms and Examples in Wittgenstein's Philosophy." In K. Ierodiakonou (ed.), *The Use of Examples in Philosophy*. Athens, Greece: Ekkremes Editions, 75–103 (in Greek).

Kindi, V. 2010. "Novelty and Revolution in Art and Science: The Connection between Kuhn and Cavell." *Perspectives on Science* 18(3): 284–310.

Kuhn, Thomas S. 1957. *The Copernican Revolution*. Cambridge, MA: Harvard University Press.

Kuhn, Thomas S. 1962, 1970. *The Structure of Scientific Revolutions*. With added "Postscript" in the second edition. Chicago: The University of Chicago Press.

Kuhn, Thomas S. 1963. "The Function of Dogma in Scientific Research." In A. Crombie (ed.), *Scientific Change Symposium on the History of Science, University of Oxford, 9–15 July 1961*. New York: Basic Books, 347–369.

Kuhn, Thomas S. 1963a. "T. S. Kuhn's Intervention in the Discussion Following 'The Function of Dogma in Scientific Research.'" In A. Crombie (ed.), *Scientific Change Symposium on the History of Science, University of Oxford, 9–15 July 1961*. New York: Basic Books, 386–395.

Kuhn, Thomas S. 1977. *The Essential Tension: Selected Studies in Scientific Tradition and Change*. Chicago: The University of Chicago Press.

Kuhn, Thomas S. 1959, 1977a. "The Essential Tension: Tradition and Innovation in Scientific Research." In *The Essential Tension: Selected Studies in Scientific Tradition and Change*. Chicago: The University of Chicago Press, 225–239.

Kuhn, Thomas S. 2000. "A Discussion with Thomas Kuhn." In *The Road Since Structure*. Edited by James Conant and John Haugeland. Chicago: The University of Chicago Press, 255–323.

Kuusela, Oskari. 2008. *The Structure against Dogmatism: Wittgenstein and the Concept of Philosophy*. Cambridge, MA: Harvard University Press.

Kyburg, Henry E. Jr. 1963. Review of *Foresight and Understanding*, by Stephen Toulmin. *The Philosophical Review* 72(1): 115–116.

Lipton, Peter. 2005. "The Medawar Lecture 2004. The Truth about Science." *Philosophical Transactions of the Royal Society* 360: 1259–1269.

Luckhardt, C. G. 1978. "Beyond Knowledge: Paradigms in Wittgenstein's Later Philosophy." *Philosophy and Phenomenological Research* 39(2): 240–252.

Masterman, M. 1970. "The Nature of a Paradigm." In Imre Lakatos and Alan Musgrave (eds.), *Criticism and the Growth of Knowledge*. Cambridge: Cambridge University Press, 59–89.

Nickles, Thomas. 1998. "Kuhn, Historical Philosophy of Science, and Case-Based Reasoning." *Configurations* 6: 51–85.

Nickles, Thomas. 2000. "Kuhnian Puzzle-Solving and Schema Theory." *Philosophy of Science* 67, Supplement. Proceedings of the 1998 Biennial Meetings of the Philosophy of Science Association. Part II Symposia Papers, 242–255.

Nickles, Thomas. 2003. "Normal Science: From Logic to Case-Based and Model-Based Reasoning." In T. Nickles (ed.), *Thomas Kuhn*. Cambridge: Cambridge University Press, 142–177.

Nickles, Thomas. This volume. "Some Puzzles about Kuhn's Exemplars."

Pickering, Andy. 1980. "Exemplars and Analogies: A Comment on Crane's Study of Kuhnian Paradigms in High Energy Physics." *Social Studies of Science* 10: 497–502.

Polanyi, Michael. 1958, 1974. *Personal Knowledge: Towards a Post-Critical Philosophy*. Chicago: University of Chicago Press.

Read, R., and W. Sharrock. This volume. "Kuhn's Fundamental Insight—Reflection on the 'Social Sciences,' as a Pedagogical and Philosophical Tool for Thinking Adequately about the Natural Sciences."

Shapere, D. 1964, 1980. "The Structure of Scientific Revolutions." Reprinted in G. Gutting (ed.), *Paradigms and Revolutions*. Notre Dame: University of Notre Dame Press, 27–38.

Toulmin, Stephen. 1953. *The Philosophy of Science: An Introduction*. London: Hautchinson's University Library.

Toulmin, Stephen. 1961. *Foresight and Understanding: An Enquiry into the Aims of Science*. Bloomington, Indiana: Indiana University Press.

Toulmin, Stephen. 1963. "S. E. Toulmin's Intervention in the Discussion Following T. S. Kuhn's 'The Function of Dogma in Scientific Research.'" In A. Crombie (ed.), *Scientific Change Symposium on the History of Science, University of Oxford, 9–15 July 1961*. New York: Basic Books, 382–384.

Toulmin, Stephen. 1970. "Does the Distinction between Normal and Revolutionary Science Hold Water?" In Imre Lakatos and Alan Musgrave (eds.), *Criticism and the Growth of Knowledge*. Cambridge: Cambridge University Press, 39–47.

Toulmin, Stephen. 1972. *Human Understanding*. Princeton, NJ: Princeton University Press.

Watling, John. 1964. Review of *Foresight and Understanding*, by Stephen Toulmin. *The British Journal for the Philosophy of Science* 15(58): 164–166.

Wisdom, J. O. 1974. "The Nature of 'Normal Science.'" In P. A. Schilpp (ed.), *The Philosophy of Karl Popper, The Library of Living Philosophers*. Vol. 14. LaSalle, IL: Open Court, 820–842.

Wittgenstein, Ludwig. 1951, 2009. *Philosophical Investigations*. Trans. G. E. M. Anscombe, M. S. Hacker, and J. Schulte. Oxford: Blackwell. (Abbreviated as PI)

Wittgenstein, L. 1958. *The Blue and Brown Books*. Ed. Rush Rhees. Oxford: Basil Blackwell. (Abbreviated as BB)

Wittgenstein, L. 1975. *Wittgenstein's Lectures on the Foundation of Mathematics*. Ed. Cora Diamond. Chicago: The University of Chicago Press. (Abbreviated as LFM)

Wittgenstein, Ludwig. 1978. *Philosophical Grammar*. Ed. R. Rhees. Berkeley: University of California Press. (Abbreviated as PG)

Wittgenstein, L. 1978. *Remarks on the Foundations of Mathematics*. Ed. G.H. von Wright, R. Rhees, and G. E. M. Anscombe. 3rd ed. Oxford: Basil Blackwell. (Abbreviated as RFM)

Wittgenstein, L. 1979. *Wittgenstein Lectures. Cambridge 1932–1935*. From the Notes of Alice Ambrose and Margaret Macdonald. Ed. Alice Ambrose. Amherst, NY: Prometheus Books.

Wittgenstein, Ludwig. 2005. *The Big Typescript*. Ed. and trans. C. Grant Luckhardt and Maximilian A. E. Aue. Oxford: Blackwell. (Abbreviated as BT)

Wray, Brad K. 2011. "Kuhn and the Discovery of Paradigms." *Philosophy of the Social Sciences* 41(3): 380–397.

ARCHIVAL MATERIAL

Letter from Thomas S. Kuhn to James B. Conant, 29 June 1961. Thomas S. Kuhn Papers, MC 240, box 25. Massachusetts Institute of Technology, Institute Archives and Special Collections, Cambridge, Massachusetts.

Letter from Thomas S. Kuhn to Donald Gillies, 2 April 1990. Thomas S. Kuhn Papers, MC 240, box 20. Massachusetts Institute of Technology, Institute Archives and Special Collections, Cambridge, Massachusetts.

6   Some Puzzles about Kuhn's Exemplars

Thomas Nickles[1]

1. INTRODUCTION

Thomas Kuhn's treatment of the role of exemplars in normal science is extremely insightful, in my view, as one of the more successful attempts to explain how science really works. Yet I find his notion of exemplar (like current talk of memes) so elusive that I am tempted to say (echoing Steven Shapin[2]): "Kuhnian exemplars do not exist, and this is a chapter about them!" My central claim will be that, insofar as Kuhn considered the key exemplars to be fixed and permanent constituents of a given instance of normal science, Kuhnian exemplars do not exist, or rarely. Kuhn's treatment of exemplars in *The Structure of Scientific Revolutions* (SSR) mirrors the strengths and weaknesses of his treatment of scientific change as a whole. The excessive rigidity of his treatment of exemplars (as I interpret him) leaves him unable to explain in any detail how a new, fundamental exemplar (as opposed to a specialized application of an existing one) comes into existence otherwise than through a sudden "aha" experience that then propagates through the individuals of the community; or how exemplars can then remain fixed within the moving frontier of normal science.

In short, I am sympathetic with those critics who find Kuhn's account of normal science too static, insufficiently evolutionary. But if we accept a broader, more flexible conception of exemplars (perhaps attributing such a view to Kuhn himself), the gap between normal and revolutionary science begins to close. For we must then recognize that exemplars can transcend scientific revolutions and that definite *lineages* of exemplars regularly do, as Kuhn himself acknowledges (12, 101–102, 189–190). (Page numbers in parentheses refer to the second, 1970, edition of SSR unless otherwise indicated.) These lineages represent a strong element of continuity through paradigm change.

When it comes to revolutionary breaks and incommensurability, Kuhn's emphasis on the failure of theory reduction is surprising since his is a problem-solving rather than theory-centered account of scientific inquiry, and since a form of *problem* reduction is central to his account of inquiry. "The unit of scientific achievement is the solved problem" (169), he informs us, and a major strength of SSR is that he tells us quite a lot about how this is accomplished in normal science. Normal scientific problems can challenge the best experts, but they are nevertheless routine in a sense. Normal scientists are able to recognize and to select for their agenda only problems that can be tamed by transforming them into well-structured puzzles and are thereby able to reduce problem solving to modeling new puzzles on old ones, the exemplars. This process involves some degree of pattern matching, using something like schema-guided, case-based reasoning rather than rule-based derivations of problem solutions "logically" from first principles. The process of direct modeling amounts to a rhetorical rather than a purely logico-mathematical reduction of a new puzzle to one or more old ones, ultimately the problem-solving exemplars that originally constituted the new paradigm.[3] Thus does Kuhn explicitly claim to relegitimize context of discovery (problem formulating and problem solving at the frontiers of research) as an epistemological topic.

Unfortunately, as Stephen Toulmin (1972) emphasized, Kuhn's overly rigid conception of normal science requires him to make an overly sharp, revolutionary break between an instance of normal science and its historical successor. And it is precisely during these revolutionary periods when the chief new exemplars presumably come into being. About how this is accomplished Kuhn says little. In a kind of regress he explains the tamed "discovery" process of *puzzle* solving in terms of a prior, more radical *problem* solving (the production of genuinely novel exemplars), but he apparently cannot explain the latter. Thus, disappointingly, he after all leaves us with an old-fashioned account (or nonaccount) of radical breakthroughs.

I shall focus on how the exemplars that make normal puzzles possible come into being, both in normal science (if they do) and in the development of a new disciplinary matrix. However, I cannot here delve deeply into Kuhn's broadly Kantian basis for answering the corresponding "How possibly?" questions: How is normal science possible? How are research puzzles possible even at the frontiers? How are exemplars possible?

Unfortunately, like his original notion of paradigm, Kuhn's talk of exemplars refracts and diffracts into a blurry spectrum of possible meanings (some already noted by Masterman 1970). For example, sometimes Kuhn presents exemplars as large theory complexes such as Newton's *Principia* and *Opticks*, Lavoisier's *Chemistry*, and Darwin's *Origin of Species* (10, 20); but at other times exemplars are smaller units, down to quite specific problem solutions, such as Galileo's treatment of pendular motion, on which other specific problems and solutions are modeled "with only minimal recourse to symbolic generalizations" ("Postscript," 189–190). Thus we can speak of *macro-* and *micro-exemplars* (my terms), both of which are exemplars, not disciplinary matrices. In what follows I shall restrict my use of "exemplar" to the micro-exemplars that Kuhn has in mind when he speaks of directly modeling one problem solution on another. We can then think of macro-exemplars, roughly, as clusters of related micro-exemplars.

It soon becomes apparent that Kuhn does not restrict exemplars to their original historical introduction. He quotes Whitehead's "a science that hesitates to forget its founders is lost" in the context of the invisibility of scientific revolutions (138), but Kuhn the internalist historian would also apply it, to some degree, to normal science. Late nineteenth-century physicists would have found neither Newton's *Principia* nor even Lagrange's *Méchanique Analytique* useful to their own work. Sometimes Kuhn's exemplars are historical tokens, including those that pop up in specialist journals; but, especially when he writes about textbooks and student education, they become types, in a sense stereotypes. In the latter sense these micro-exemplars are "the sorts of standard examples of solved problems which scientists encounter first in student laboratories, in the problems at the ends of chapters in science texts, and on examinations" (Kuhn 1970b, 272).

The items that Kuhn labels exemplars are diverse in other respects as well. Sometimes "exemplar" seems to refer to what we today call idealized models, such as the idealized point-pendulum, but sometimes the type of physical motion itself and not only its abstract representation must be included among the exemplars.[4] SSR also prominently features the Leyden jar and Atwood's machine as exemplary. Given his emphasis on scientific practices and their conveyance from master to apprentice, he wants to include "law, theory, application, and instrumentation together" (10), yet he also says that modeling on exemplars is prior to rules and other symbolic generalizations, that an exemplar is "a unit that cannot be fully reduced to logically atomic components" such as "concepts, laws, theories, and points of view that may be abstracted from it" (11; Chapter 5). At the very least, then, exemplars can include more than symbolized problem solutions. Kuhn also speaks of a "standard set of methods or of phenomena" (13). Had he given more attention to biological research, he probably would have included model organisms such as the roundworm *C. elegans* and the fruit fly, *Drosophila melanogaster* (Love 2009). I contend that he should have included exemplary *bad* practices, "negative exemplars" as well, since they play a role in both teaching and research. Given the importance to his epistemology of Kuhn's quasi-Kantian cognitive theory, with its "acquired similarity relation," there is also the question of whether exemplars are to be represented instead by some cognitive structure such as a schema or a set of basic cases in a case library (Nickles 2000, 2003b; Nersessian 2003; Bird 2008; but see Kindi, this volume).

Despite this flexibility, some passages in Kuhn's account of normal science strongly suggest that exemplars, once stabilized, remain fixed in essential respects that allow for only superficial changes in formulation. At the very least it remains unclear at what level of description the very same exemplars are the constitutive features of a long stretch of normal science. For example, Kuhn speaks of normal science as seeking no essentially new phenomena or theory but as "a strenuous and devoted attempt to force nature into the conceptual boxes supplied by professional education" (5) and of attempting "to force nature into the preformed and relatively inflexible box that the paradigm supplies" (24). "Even the project whose goal is paradigm articulation does not aim at the *unexpected* novelty" (35; emphasis in original). As for the achievement of the solved problem, in normal science "at least part of that achievement always proves to be permanent" (25).

I end this section on a positive note. By stressing the importance of direct modeling and the cognitive skills required to master it, Kuhn integrated rhetoric (analogy, metaphor, simile) with logic and mathematics in a way that the then-dominant philosophy of science had usually forbidden. This was key to his account of problem solving by direct modeling. Like Kant against the British empiricists, he realized that what was previously taken as given and transparent—scientists' ability to recognize good problems and solutions, interesting phenomena, etc., and the relevant rhetorical relations among them—requires an elaborate, socially induced cognitive structure to account for its possibility. Moreover, Kuhn's emphasis on modeling anticipated today's concern with models and modeling practices and, thereby, the decentering of theories in favor of more emphasis on expert practices and on scientists' motivations. In so doing, Kuhn attempted not only to bring context of discovery[5] back into his model of scientific change, where, strangely enough, it had previously been treated as an exogenous factor, but also to demote standard confirmation theory. In his view no neutral, formal confirmation theory can fit real science, since good judgment depends heavily upon content-laden expertise. According to Kuhn not even an expert from a different scientific field, let alone a logician or philosopher of science, is competent to make normal-scientific judgments (3–4, 49; Hoyningen-Huene 1993, 138). His frequent mention of heuristic promise, as judged by experts, is central to his positive account of both normal and crisis science.

2. NORMAL AND REVOLUTIONARY DYNAMICS

According to an old pragmatic dictum, attributed to Ralph Barton Perry, there are two ways to solve a problem: you can either get what you want or you can want what you get (Reitman 1964, 308). These two ways correspond pretty nearly to what Kuhn in SSR termed, respectively, normal science and extraordinary science. Most outcomes in normal science, Kuhn tells us, are highly anticipated, the challenge being how to get to those results from extant exemplars (36). Even the "how" is so constrained that normal scientists know in advance the kind of thing that is required. By contrast, the breakthroughs that become candidates for resolving a crisis in terms of a new approach often result from unorthodox and hence unjustified moves. Here the target may be known in advance (e.g., getting the Balmer and other spectral series), but the known kinds of "how" have failed and may even be deemed incompatible with the needed result (as with the specific heats and the ultraviolet catastrophe problems). In both the latter cases it appears that a conceptual breakthrough of an unknown kind may be necessary, one that directly violates old assumptions or at least one for which the set of exemplars that constitute the old paradigm is seriously inadequate. Often enough, the first steps involve an unorthodox move that yields a promising result, but it remains unclear precisely why it does so (e.g., Planck's quasi-Boltzmannian trick that enabled him to "derive" the black-body radiation law in 1900 and Debye's later *Kunstgriff* in extending quantization to the anharmonic oscillator). If the new approach takes hold, subsequent normal scientists, often over generations (as was the case in understanding the implications of Planck's law), attempt to reverse engineer the breakthrough in order to distill out its "essence." A more thoroughgoing understanding of why it works in the initial cases can be crucial to extending it to new ones (as in Ehrenfest's adiabatic principle).

During extraordinary periods, venturesome Kuhnian scientists adopt an epistemic attitude that is doubly pragmatic. In addition to "the unorthodox new approach seems to work well enough to extricate us from this problem situation," there is the equally important "and we see some good ways in which *we can continue to use it in future research.* It opens up new spaces for research, thus guaranteeing the continuation of the enterprise." A radical new approach thereby gets purchase. It makes a pragmatic difference, a difference to future scientific practice. Here we encounter the form of evaluation that I term *heuristic appraisal*, of which more below.

I want to maintain that, to a lesser degree, this pragmatic description also holds of the more creative *normal* scientists. Kuhn speaks of scientists having to "beat nature into line" (135). Kuhnian normal scientists have to beat their own results into the line of their paradigm in order to make them fit existing genealogies of exemplars, sometimes by the sort of post hoc rationalization that is more visible in revolutionary periods. Only thus (I claim) could physicists beat the new statistical physics and electromagnetic theory (with its noncentral forces, advanced and retarded potentials, etc.) into the "Newtonian physics" genealogical lines. The puzzle pieces must be *made* to fit. Sociologically, this point means keeping the *practitioners* in line, that is, solving the problem of social order for these communities, a topic fruitfully discussed by Barry Barnes (2003).

What worries me is that Kuhn plays both ways with this sort of example. Beating nature into line (135) is one mechanism for preserving the robustness of the paradigm, but so is beating a theoretical innovation into line with "nature." It would seem that Kuhn employs the same kind of example to demonstrate a *loosening* of the paradigm during a crisis, thus blurring the distinction between normal science and revolution by allowing degrees of (sometimes permanent) relaxation of the reigning paradigm.

What are we to do with Maxwell's electromagnetic theory, for example? Kuhn himself speaks of it as revolutionary (7) and even as effecting some degree of paradigm change (66), yet he, like most writers, regards it as a solid contribution to classical mechanics. His defense is that, first, like the wave theory of light, it was a relatively localized and "professional" paradigm shift (or so he claims). Localized compared to what? Kuhn presumably means that it had only a small reach beyond the physics community, unlike the Copernican, Newtonian, Darwinian, and Einsteinean revolutions. (However, requiring such reach creates trouble for much else that Kuhn says about revolutions; e.g., he later allows for revolutions in small subspecialties.) Second, Kuhn notes that Maxwell himself adopted a Newtonian mechanistic approach to modeling electromagnetic phenomena in the ether, an approach that Maxwell and others presumed would eventually be "compatible with some articulation of the Newtonian mechanical view" (74). For Kuhn revolutions are events of which the participants are actively aware as such. Third, there was indeed a revolution, Kuhn says, but it came only later as trouble mounted for the optimistic view that the paradigm could account for the interactions of matter and ether, thereby creating the crisis that led to Lorentz, Poincaré, and Einstein (plus, we could add, the emergence of symmetry and invariance requirements as possible new components of the resulting disciplinary matrix).

Kuhn in effect accommodates the Maxwell case by allowing that paradigms can change significantly in a noncumulative manner, yet without revolution. He hints that the same could be said about the wave theory of light earlier in the century, and he would surely have to say the same about the emergence of thermodynamics and statistical mechanics. Regarding thermodynamics, after all, in their *Treatise on Natural Philosophy* (1867) William Thomson and Peter Guthrie Tait had replaced Newtonian force by the new concept of energy as the central concept of physics.

Kuhn's defense suggests that Maxwell's work looks revolutionary only by hindsight, that it would be Whiggish to read back into Maxwell's generation recognition of the the conflict in invariance principles later highlighted by Einstein. This response is largely correct, but I remain uncomfortable with Kuhn's treatment of the Maxwell case. True, Maxwell's work, in its time, was certainly not as revolutionary as Einstein's. My point, however, is that it was iconoclastic enough to produce a significant change in the received paradigm. The admission and mathematization of non-central forces (for example) was a serious departure from the Newtonian paradigm as the latter existed prior to researches into magnetism, as Kuhn acknowledges. So what happened to Kuhn's claim that normal science is cumulative? Couldn't an analyst just as easily say that the historical changes in so-called classical mechanics were collectively transformative rather than additive, that the difference between what Kuhn labels normal science and revolutionary science is really only one of degree—mainly one of speed or time scale?

I claim that such transformations occur also in the genealogies of individual exemplars that persist, at least in name, over time. Think again of the harmonic oscillator and of the changing status of Planck's alleged quantum breakthrough of 1900, in Kuhn's own account of it in his history of the early quantum theory. On a different note, in "Postscript-1969" he had mentioned changing professional affiliations and citation patterns as marks of revolution, but he would surely have to admit that the patterns change, albeit more slowly, during normal science as well.

Kuhn stresses continuity and gradualism when he wants to, especially when he takes the scientists' point of view or employs his evolutionary analogies. When taking the philosopher's point of view, however, he stresses discontinuity and revolutionary breaks. Thus special relativity fails to reduce classical mechanics, given his philosophical assumptions about meaning (101f); whereas many physcists at that time and later readily regarded special relativity as reducing to classical mechanics in the range of low velocities, and, they now, post-quantum mechanics, consider special relativity to be a largely "classical" theory. Yet when we consider Maxwell's theory from the philosopher's point of view, it is not strictly compatible with a pre-maxwellian conception of Newtonian mechanics either. It was far from the simple addition of a special force law to the Newtonian system. Since the labels "classical mechanics" and "classical physics" themselves did not come into being until after the twentieth-century revolutions, one may ask if Kuhn himself is not being Whiggish in lumping so much under a single, overarching, classical mechanics paradigm. At the very least he has to acknowledge that major new exemplars originate within normal science as he here characterizes it (but see next section).

Should a Kuhnian reply that, at a sufficiently abstract level, physicists maintained the integrity of "classical" mechanics by attempting to figure out a way to absorb electromagnetic theory (and statistical mechanics, etc.) into its framework, my response would be that such modes of absorption are rather similar to the kinds of moves that Kuhn himself ascribes to successful revolutionaries who then stress the continuity of their work with what went before. As a historian Kuhn himself was acutely aware that scientific work from Descartes, Huyghens, and Newton to Faraday, Maxwell, Kelvin, and Boltzmann transform the very concept of mechanics almost out of recognition. Some historians (e.g., Brush 1988) speak of a second scientific revolution (in a big sense of 'revolution') in the century after Newton, with the emergence and maturation of the "Baconian" sciences such as electricity and magnetism. And historians largely agree that modern mathematical physics did not emerge until after the turn of the nineteenth century in France, or even later in Britain and Germany.

In my view Kuhn's treatment of classical mechanics and other long-term paradigms lands him in a dilemma. Either he must allow that normal science is more dynamic than his (usually) rigid account permits, or else he must break up long-term endeavors such as "classical mechanics" into a series of different paradigms with revolutions between them that even Kuhn himself renders invisible. His account of the three types of puzzles that normal scientists engage is simply *too* tame.

More strongly still, I claim that the exemplars themselves evolve during a period of normal science, in two respects: (1) the set of key exemplars changes by both addition and subtraction in ways more profound than by Kuhnian articulation; and (2) individual exemplars themselves evolve in time.[6] Although more obvious in the case of alleged long-term paradigms such as classical mechanics or "Darwinian" evolutionary theory, I believe that this will be true even of relatively short-lived paradigms, given the extensive amount of reverse-engineering, multi-pass work necessary to construct them as mature paradigms—and that this process cannot be confined to an initial "extraordinary research" period (Nickles 1997).

My defense of these claims must remain sketchy. The attempt to model Kuhnian normal science in the next section is part of it. For now a theoretical argument is that normal science satisfies the conditions for evolution: variation, selection, and retention or transmission within a relatively steady social and intellectual environment with selection pressures. Thus we should expect to get genealogies of exemplars, the splitting of exemplars into more specialized ones that largely replace the original in practice (188–189), symbiotic conjunctions of exemplars that become practical units in their own right, and so on.[7] Then there is the historical evidence. Here are a few more examples.

Kuhn claims that Galileo saw a pendulum bob as attempting to reach its previous height, whereas Aristotle saw it as straining to reach the center of the earth. Until the scholastics invented the impetus theory in which Galileo was educated, Kuhn writes, "there were no pendulums, but only swinging stones, for the scientist to see. Pendulums were brought into existence by something very like a paradigm-induced gestalt switch" (120). But surely today's physicists regard Galileo's pendulum in a quite different way than Galileo did, namely as an approximation to a simple harmonic oscillator, thus as a special case of a general phenomenon (harmonic oscillations) that is, in turn, a special case of a still more general phenomenon (oscillations in general). Here I have in mind a series of developments, including Hooke's law, later mathematical work on sine functions and their graphs, Fourier analysis, Maxwell's electromagnetic theory, and so on up to string theory, not to mention the need to correct Galileo's own account of pendular motion. Modern physicists see comparable aspects of the world (and engage it practically) in a way quite different from Galileo. Kuhn knows this, of course, but these facts do not sit well with his tame presentation of normal science. The case becomes all the more convincing if we include in the notion of exemplary standardized expert practices, experimental designs, instrumentation, model organisms, and the like. Think of the microscope as an exemplar. The changes in all of these, taken together over a productive stretch of normal science, will transform community life.

Just as new instruments, practices, and conceptual exemplars can come into existence in an allegedly normal science, they can also disappear. Sometimes they are deemed simply wrong. Often they are displaced by more advanced equipment, more sophisticated experimental designs, mathematical techniques, and the like. In short, Kuhn's normal science is simply *too* cumulative. The topic requires further discussion than I have space for here.

Kuhn himself says that science is structured, somewhat like biological populations, into a pattern of lineages of puzzles and solutions. His most extended example occurs in the "Postscript." Huygens modeled the physical pendulum on an organized collection of Galileo's simple pendulums, and, some decades later, Daniel Bernoulli was surprisingly able to model the flow of water from an orifice on Huygens' physical pendulum. The obvious difference from a biological lineage is that problem-solving lineages have a different topology, forming more of a crisscrossing network of lineages, since scientists may solve a typical puzzle by drawing on several exemplars as its "parents." Like other examples of human technology, a problem solution is likely to have a position in more than one lineage.[8]

What is interesting is that these lineages are apparent not only within normal science but also through scientific revolutions. And they immediately suggest at least a partial answer to the question, Where do exemplars come from?, one that obviates the need to talk of revolutionary breaks in such an all-or-nothing manner. It was not just the equation of human perception with scientific observation that led Kuhn to speak of gestalt switches. It was also that his treatment of the appearance of radical new exemplars remained too much in the "aha" tradition, in contrast to his own detailed work on Bohr's 1913 atom model with John Heilbron (Heilbron and Kuhn 1969) and on the emergence of the old quantum theory (Kuhn 1978), accounts in which sharp revolutionary breaks are no longer apparent.

The obvious way to explain the existence of genealogies is in terms of an evolutionary mechanism, in this case one that is selectionist but not otherwise directly analogous to biological evolution. Human communities do have to build on the resources available to them, after all, and the primary way is by modifying and recombining forms already available.

3. FOUR MODELS OF NORMAL SCIENCE

How static is normal science? Where do its exemplars initially come from? Can essentially new exemplars emerge during the course of normal science, or would this challenge the robustness of normal science, as Kuhn understands it? Can the basic exemplars evolve in important ways during the course of normal science? In this section I set out four models of normal science, four possible interpretations of Kuhn's words. I have been unable to find a model fully consistent with what I take to be plausible interpretations of Kuhn's various statements. All four of the following capture something of Kuhn, but not everything; and all are problematic.

Model 1 assumes that (a) the disciplinary matrix governing normal science is largely constituted by its most basic exemplars (taking seriously Kuhn's idea that both symbolic generalizations and values are embodied in the exemplars as ideals and largely abstracted from the latter); (b) these basic exemplars therefore must be present from the beginning of genuinely normal science; and (c) they change very little during the course of that instance of normal science. This model results from a strong interpretation of what Kuhn means when he speaks of the paradigm as *guaranteeing* the solution of any puzzle to which it legitimately gives rise. He writes, for example:

> Perhaps the most striking feature of the normal research problems we have just encountered is how little they aim to produce major novelties, conceptual or phenomenal. Sometimes, as in a wave-length measurement, everything but the most esoteric detail of the result is known in advance, and the typical latitude of expectation is only somewhat wider. (35)

> Even the project whose goal is paradigm articulation does not aim at the *unexpected* novelty. (35; emphasis in original)

> Though intrinsic value is no criterion for a puzzle, the assured existence of a solution is. (37)

Model 1 interprets Kuhn as saying that normal scientists behave as though there is a set of *basis exemplars* that span the problem space of that paradigm in the sense that every legitimate research puzzle is solvable in terms of modeling on one or more members of this set of exemplars. Here my analogy is to the set of "basis vectors" in linear algebra, vectors that span a vector space in the sense that every vector in the space is equivalent to an algebraic sum of the basis vectors. Kuhn's own, less formal analogy is to a legal system in which the published precedents, together with the law itself, are considered a sufficient basis to resolve all normal legal disputes.

Accordingly, the paradigm promises that the new puzzle and its solution can be fit into the genealogical network that characterizes that research tradition. And this explains why normal scientists cannot, as such, seek essential innovation, something that would introduce a new basis exemplar, either de novo or by transforming one already in place (Hoyningen-Huene 1993, 174ff.).

Model 1 also assumes that the exemplars do not appreciably change from the point of their first introduction. In Chapter 11 of SSR, "The Invisibility of Revolutions," Kuhn notes that it is standard textbooks that invariably stabilize normal science, by reproducing the standard exemplars. He continues:

> Textbooks, however, being pedagogic vehicles for the perpetuation of normal science, have to be rewritten in whole or in part whenever the

language, problem-structure, or standards of normal science change. In short, they have to be rewritten in the aftermath of each scientific revolution, and, once rewritten, they inevitably disguise not only the role but the very existence of the revolutions that produced them. (137)

Partly by selection and partly by distortion, the scientists of earlier ages are implicitly represented as having worked upon the same set of fixed problems and in accordance with the same set of fixed canons that the most recent revolution in scientific theory and method has made seem scientific. No wonder that textbooks and the historical tradition they imply have to be rewritten after each scientific revolution. And, no wonder that, as they are rewritten, science once again comes to seem largely cumulative. (138)

Notice that Kuhn sees significant change in exemplars only from one revolution to the next. We might wish that he had made a study of changes in textbooks during the long period of the retrospectively labeled "classical mechanics," which, as we have seen, he seems to treat unhistorically as one long period of normal science.[9] For the fact that the texts *can* be rewritten in such a manner signals that interpretation is in play, hence that subtle but significant shifts in interpretation can occur as normal science proceeds. As a matter of historical fact, we know that the textbooks do get rewritten, usually to a significant degree, not only during revolutions but also during sufficiently long periods of normal science.

Obviously, then, Model 1 immediately faces severe difficulties, both as an account of science and as an interpretation of Kuhn, given other things that he says. (a) It is historically mistaken to think that all of the basis exemplars are already present at the birth of normal science under a new paradigm, having been generated quickly during the revolutionary interregnum. Some of Kuhn's own examples (e.g., Maxwell's statistical and electromagnetic theories as contributions to classical mechanics) refute this interpretation. A major new paradigm (supposing such to exist) never springs fully formed from a crisis. Rather, there is a highly constructive transitional period to which Kuhn gives insufficient attention in SSR, with the result that it is unclear when we should say that normal science proper begins. He did provide such attention in his detailed history of the early quantum theory (Kuhn 1978), but at the cost of losing his sharp normal-revolutionary distinction, as critics have complained (Klein et al. 1979).

(b) The all-or-nothing character of Model 1 leaves us with a dilemma: either normal science is characterized by a preexisting set of basis exemplars or not. If it is, then the onset of normal science is far more abrupt, and normal science itself is a far more static affair than the history of science allows: all the exemplars are present from the beginning to the end, and they do not evolve. But if normal science is not characterized by such a basis set, then normal scientific work itself generates new basis exemplars (not merely specializations of already extant exemplars), that is, major innovations—meaning that normal scientists themselves are engaged in constitutive paradigm creation and not only in the kinds of articulation that Kuhn identifies. Assuming that the new basis exemplars are coherent with those already in place, they are likely to contradict aspects of the previous paradigm, thus blurring the distinction between revolutionary science and normal science. Furthermore, a new basis exemplar would provide a disruption at least as serious as an unexpected empirical discovery, for both are anomalous challenges to the presumed overall completeness of the paradigm. It would seemingly amount to a minirevolution. The difference between revolutionary and normal science would reduce to one of time-rate of innovation.

(c) In making normal science an all-or-nothing affair, Model 1 also undermines Kuhn's account of scientists' motivation to work in those areas that have high heuristic promise. For on this model, science becomes fully normal only at that point when it has exhausted its major creative potential. In this regard, Kuhn himself once slipped into the ill-advised phrase of "mop-up work" (24). Complete maturity is the point at which highly creative thinkers either switch to more challenging problems, perhaps in other fields, or generate a crisis by now having the luxury to focus attention on the most miniscule and esoteric discrepancies. They can thereby elevate anomalies—like Lord Kelvin's "two small clouds on the horizon" (Kelvin 1901, 1)—into major problems and proceed to address them in novel ways. Insofar as perceived sterility or more interesting futures are the case, it is not only anomalies but also their very *absence* (i.e., the absence of challenging research problems) that can trigger a crisis for the self-conception of the field-cum-community of experts, for the field is equally in danger if there is nothing important left to do.

(d) I claim on historical, cognitive, frontier-epistemological, and evolutionary grounds (the conditions for evolution mentioned in §2) that even the basis exemplars will change in character over time during normal science. As new puzzles are morphed to fit classical exemplars (or some combination of them), the exemplars themselves undergo a series of (often gradual and subtle) reinterpretations. In the more challenging and creative cases, it is a question of *mutual* fit between the exemplars and puzzle solutions. Exemplars are not fixed centers around which research revolves. To change the metaphor, there are interaction terms in the dynamics of normal science that Kuhn does not take into account. As noted above, not only nature but also the paradigm must be "beat into line." A general point about the diffusion or application of techniques to new contexts is that they typically change in the process.[10] Kuhn's Wittgensteinian influence could have helped him here, as I should like to say of exemplars that *their meaning is given by their use* (cf. Sharrock and Read 2002; Kindi, this volume). In fact, Hoyningen-Huene (1993, 154ff.) suggests that this was Kuhn's own view. Hence, as they are applied to new puzzles, we should expect their meaning to shift.

Kuhn provides no very good way to identify and individuate exemplars. He writes that it is the business of normal science to articulate the paradigm, extending it to new areas and filling in gaps, but such passages are ambiguous between paradigm as disciplinary matrix and paradigm as exemplar. At any rate I want to claim that normal science also articulates the *individual* micro-exemplars in various ways, including finding ways to combine two originally distinct exemplars as particularizations of a more general one. Consider again Galileo on pendular motion and think of the later development of an elaborate body of results concerning simple harmonic motion and variations such as damped, driven, and coupled oscillation, not to mention work such as Fourier series. Or think simply of Galileo on falling bodies without a developed conception of gravitation.

Another route to a similar conclusion is this. Given Kuhn's correct emphasis on the importance of metaphorical language at the frontier of research, one aspect of paradigm articulation will be that newly introduced metaphors such as "force," "gravity," "energy," and "magnetic field" must gradually become dead metaphors with a precise, literal, technical meaning as the normal science matures. Yet the direct modeling insight produces an opposite conclusion in another respect in that, being models, even the old basis metaphors *regain rhetorical life* as analogies, similes, and metaphors.[11]

Thus we arrive at the idea that normal science itself has a deeper history than Model 1 suggests. Insofar as this model captures Kuhn's own view, this is surprising given his general emphasis on the importance of interpretation and on the heightened sense of interpretive nuance that the then-new history of science brings. Kuhn's account of revolution and incommensurability manifests an amplified sensitivity, while his account of normal science suppresses it. After all, even normal scientists are working at their respective frontiers and must deal, admittedly on a smaller scale, with the kinds of uncertainties and risks that revolutionary scientists do.

Model 2 is a variation of Model 1 that is more explicitly dynamic in recognizing that the reigning paradigm does not guarantee that all puzzles legitimately formulated in its terms can be solved *directly* in terms of the original basis exemplars. Rather, it promises that it has the resources to *generate* the new, specialized exemplars that may be needed for efficient work, some of which may become basis exemplars at the new level of esoteric detail as a new subspecialty emerges. Kuhn's treatment in "Postscript" suggests such a view (188–189). There he treats basic principles such as $F = ma$ as schematic rather than content-rich laws, arguing that much of the empirical content enters through specialized equations such as those for coupled oscillators. So does the pendulum-efflux lineage, in which Kuhn regards normal science as becoming steadily more specialized. Kuhn gives us a hierarchical picture of a large, overarching paradigm with smaller paradigms (the smallest with perhaps twenty-five practitioners) nested within it.

But Model 2 does not escape the dilemma, for it retains Model 1's idea that all exemplars within a given normal science derive from an original set of basis exemplars that do not change, either by addition or by evolutionary transformation.

Model 3 asserts that *omne exemplar ex exemplari*: all exemplars, including the most basic ones, derive in part, and in the direct-modeling sense, from one or more prior exemplars—sometimes across revolutionary divides. It gives up the claim that there is a new set of basis exemplars, present from the beginning. It explicitly extends to all exemplars (exemplary problem solutions) Kuhn's insight about the direct modeling of new problem solutions on old ones. After all, in one of his more definitive statements, Kuhn himself regarded revolutions as transformations or reorganizations of materials already available for the most part.

> Just because it did not involve the introduction of additional objects or concepts, the transition from Newtonian to Einsteinian mechanics illustrates with particular clarity the scientific revolution as a displacement of the conceptual network through which scientists view the world. (102)

An attractive feature of this model is that it provides at least a schema for understanding how creative scientists achieve innovations in extraordinary science, namely, by extending (somewhat relaxing) Kuhn's insights about puzzle solving in normal science. Thus the accomplishments of revolutionaries no longer drop out of thin air. But Model 3 surely remains too simple and too conservative in supposing that all major creative episodes can be understood in this way. It appears to start a regress back to a set of ancient, underived, and hence unexplained exemplars, as if the entire history of science consists of footnotes to Plato, so to speak.

This last criticism is not entirely fair, however, for the idea of derivation here in play is a flexible one. Model 3 involves *partial* derivations based on cognitive modeling, not necessarily strict logical derivations or even ones involving smooth mathematical transformations. Think of Galileo, Descartes, and Newton pondering Aristotle's view that a constant push is needed to keep nonnatural motion going, as they began to articulate the modern principle of inertia. So it is plausible to think that the regress eventually dissolves into quite ordinary assumptions made by people in everyday life.

Model 4 drops the assumption of the previous models, that all of the central exemplars of a successor normal science are somewhat new to it, and adds two others. (a) Some exemplars may be simply carried over from the old paradigm with no significant change; (b) others may be carried over in a modified manner via what we might call a new *metaparadigm*; and (c) still others may be introduced from another field, as in Kuhn's early discussion of Sadi Carnot's work (Kuhn 1960; Barnes 1982, Chapter 1). (b) recognizes the kinds of intertheory relations suggested by the transition from classical mechanics to special relativity, in which many classical equations can be modified to take into account the velocity of light *in vacuo* as a limit velocity, by appropriately inserting the Lorentz factor $(1-v^2/c^2)^{1/2}$. And somewhat similarly for the transition to quantum theory, in which suitable quantum conditions are imposed on systems inherited from classical mechanics.

Invoking his holistic conception of meaning, Kuhn balks at regarding such transitions as genuinely cumulative theory reductions (100ff., 149–150), but he is alive to the importance of heuristic devices such as Bohr's correspondence principle in its various forms and Ehrenfest's adiabatic principle and (later) theorem. It is no accident that successor theory systems tend to resemble their predecessors in significant respects. In the "Postscript," Kuhn writes:

> Since new paradigms are born from old ones, they ordinarily incorporate much of the vocabulary and apparatus, both conceptual and manipulative, that the traditional paradigm had previously employed. But they seldom employ these borrowed elements in quite the traditional way. (149)

Yet he was also well aware that, at the frontiers of research, creative scientists are *not* so concerned about the delicate meaning shifts on which historians and philosophers may focus as to jettison unorthodox problem-solving devices that seem to them to yield good results. Quantized oscillators are still oscillators.

Model 4 is not without problems, even as a better interpretation of Kuhn than Models 1–3. It retains the original problem of fixity of exemplars within normal science. And, *pace* Kuhn's concern about meaning change, like Model 3 it implies a fairly strong *continuity through revolutions*. Besides the straight carryovers added in Model 4, there are the lineage continuities of Model 3, that is, lineage continuities of key problems and solutions that transcend Kuhn's revolutionary breaks. So, once again, revolutions turn out to be less radical than Kuhn claims. Just as the paradigm-producing revolutions that found a new science are constructed from resources already available, as historians and sociologists of science have been right to emphasize, so are the paradigm-changing revolutions within an already mature science.

There are pressures from within Kuhn's own views about science to recognize more continuity than his talk of incommensurability allows, both from his own perspective as a former physicist, now historian and philosophical modeler, and from the perspective that he attributes to his scientific practitioners. First, given his emphasis on science as a problem-solving activity that produces lineages of problems/solutions, in place of the old theory-centered tradition, it is surprising that he makes so much of the failure of a theory to reduce its predecessor, when there exist prominent problem lineages that tunnel right through the alleged revolution barrier with its supposed incommensurability. Second, for all of the "postmodern" implications of some of his views, Kuhn himself clearly remains in the Enlightenment-Modern tradition of regarding mature science as a very special and important social enterprise, one that progresses in more evident ways than any other human endeavor—including progress through revolutions (160ff.). Thus he cannot afford to say that a revolution destroys the integrity of a scientific field by changing the subject. Third, we come to the motivation of Kuhn's creative scientific practitioners. The end of their field would be their end as creative scientists. In order to maintain their identity as experts across paradigm shifts, iconoclastic scientists must argue that their new approach retains key elements of the prior paradigm, in redescribed form. Kuhnian crises are so worrisome because it appears that a once-productive field has come to a dead end and, consequently, that one's acquired expertise and reputation will no longer be productive and respected, worse, that one's life work will have been in vain. All the proposed alternatives that may lead somewhere seem (at least to those who are veterans in the field) to change the subject or at least to violate one or more defining conditions, or conditions of intelligibility, of the previous scientific game. Thus the field's practitioners strive mightily to preserve whatever semblance of continuity they can. Even when a new approach gains control of the field, post hoc rationalizations come thick and fast. This is not merely a case of the victors rewriting history to justify their "rational" continuation of the field or to celebrate their victory at the expense of their opponents. It is at least as much an attempt to maintain a sense of self-identity by "saving" the field as an active research site. Meanwhile, it is not uncommon for the practitioners of a losing research program to redescribe their own work in a manner that makes it look anticipatory of the eventual change.

Those who make the revolutionary transition see themselves as maintaining the integrity of the field, even as having saved it from serious danger or having purified it (Kindi 2010). Thus it is important that the new textbooks reveal the revitalized scientific domain as a continuation of what went before for both sociopsychological and technical-intellectual reasons of professional expertise. A text that presented the new work as totally different (say, as an attempt to legitimize a mere power play) could scarcely get off the ground, and recruitment of new students would fail. New students must still be educated on the most recent version of the old exemplars, from which the new and more technically sophisticated ones loosely derive.

4. WHERE, THEN, DO EXEMPLARS COME FROM? WHAT MAKES AN EXEMPLAR?

Section 3 raised but did not fully answer the question of where exemplars come from. We now have a somewhat better grip on where Kuhnian exemplars as entities come from but not yet what makes them exemplary, that is, ideals to emulate and thus to employ as guides and as tools in further work. What makes something an exemplar, a problem-cum-solution of the sort that is selected for inclusion in a textbook, widely cited by experts in the field, or the design of an instrument or a technique? What makes some problems rather than others the central problems of the field? What makes a problem-cum-solution worth selecting for inclusion in an annual review of the literature or in a textbook? How does a piece of equipment or new experimental design become recognized as standard? Seeing exemplars as part of lineages provides part of the answer, in that an exemplar in a new normal science often, even usually, inherits some degree of exemplary status from its predecessor. Especially insofar as it alters a previous exemplar, of course, it must prove its mettle in the new context. Any brand new exemplar-candidates are initially in a state of even greater probation.

Here Kuhn points us in exactly the right direction, I believe, although in SSR he provides little detail about how the process unfolds, surely because that is a compact work and the story, both sociohistoriographically and philosophically, is going to be complex, even messy. (Again, Kuhn provides a good bit of detail in his 1978 history of early quantum theory.) The direction in question is toward the future. SSR is noteworthy among the literature of the time for its emphasis on heuristic fertility and future promise, a complex of issues that I lump together as the role of *heuristic appraisal* in scientists' judgments about what to do next and which available items to employ in their own work.[12] Of course, Kuhn the internalist historian did not give sufficient attention to the changing social environments, including funding opportunities, in which scientific communities exist.

There are many dimensions to heuristic appraisal and, so, many possible factors that potentially figure in such judgments. In SSR Kuhn provides two major reasons when he tells us that a paradigm is an unprecedented achievement that attracts an enduring group of practitioners but one that is "sufficiently open-ended to leave all sorts of problems for the redefined group of practitioners to resolve" (10). This characterization has a retrospective component and a prospective component in which promise of future fertility is paramount.[13] Kuhn expresses the point in Aristotelian terms: "Normal science consists in the actualization of that promise . . ." (24). At this point in SSR Kuhn is still using the term "paradigm" ambiguously, at least between macro- and micro-exemplars, but I suggest that his point applies to every exemplar, especially the most basic ones.

The retrospective component is familiar. An exemplar candidate, like any tool, will gain status if it shows itself useful in a variety of related situations. In a sense scientists vote by positive and negative citation of previous work (another sort of network that, to some degree, crosses revolutionary boundaries), with the "vote" of highly respected figures counting for more. Frequently cited work is more likely to be presented in future textbooks, in simplified and idealized form, as something useful for students and practitioners in neighboring specialties to know. In this respect, exemplars emerge as the product of a kind of specialized crowd sourcing by the normal scientific community. (The crowd-sourcing image is my own, not Kuhn's, but it brings out his point that justification is, at bottom, social, that there is no authority higher than that of the expert community.) For Kuhnian normal science more than for revolutionary science, "widely useful in the past" implies "likely to be useful in the future." A successful track record leads normal scientists to believe they have the correct answers, techniques, etc., solid platforms on which to build.

This brings us explicitly to the other half of Kuhn's answer: future promise, which we can regard as a criticism of standard confirmation theory. For something to gain the status of an exemplar multiple scientists have to judge it a valuable guide to their planned future work. Being models to emulate, exemplars possess a forward-looking, normative component. Past success as a sign of correctness, truth, or high probability is not a sufficient explanation of this attraction for future action, for three reasons. Traditional confirmation theory tends to stop here—with degrees of belief or something similar as the future-oriented output. But holding true (or nearly enough) in the future is not really what matters (or is not all that matters) to Kuhnian practitioners. They want to be able to see how they can *use* these tools to advance their own work, to see what they can *do* with them. As we all know, there is a big difference between knowing and teaching something, or merely holding it as part of one's worldview, and seeing how to use it in ongoing, creative research. Second and related (and as noted above), Kuhn denied that any content-neutral formal theory could capture expert scientific decision making. Only the relevant experts are equipped to see the future possibilities at all clearly (which is not to deny that their claims can be challenged by external reviewers of various kinds). This point still holds under the Kuhnian shift from a theory-centered to a problem-centered account of science. For Kuhn a paradigm debate is not only about past problem-solving successes and failures. Creative scientists are willing to tolerate "Kuhn loss" (as others have dubbed the problem solutions that no longer hold up under the new paradigm), provided that that the new approach promises to break through a perceived impasse or to open the door to a new world of research delights.

Third, among other things, confirmation theory leaves us with the problem of new paradigms or theories, to which Kuhn perhaps first called attention: why would serious scientists, with their reputations and careers on the line, abandon a highly accomplished paradigm in favor of a badly undeveloped, unorthodox, even crazy-looking new approach with a slim record of success? Because, in Kuhn's view, in cases of crisis perceived heuristic fertility can outweigh an impressive track record for those scientists who believe that the old approach is either defective or largely played-out. This is certainly not to deny the utility of statistical methods of hypothesis testing at the local level of normal research.[14]

The role of heuristic appraisal is similar in the muted context of normal science and for individual exemplars. The fact that scientists refuse to reject normal-science supporting principles in the face of anomalies indicates how strong a motive for choice heuristic appraisal can be, in Kuhn's view. Perceived heuristic fertility can be strong enough to overcome what confirmation theorists would label a negative track record. As for individual exemplars, nothing that lacks an aura of heuristic fertility can become an exemplar.

Taking the point of view of the professional scientist, Kuhn understands the strong attraction that actual and potential exemplars possess. A research scientist can retain that self-image and identity within the scientific community only as long as s/he is actively working on research problems. Scientists adopt something as exemplary because they think they can *do something interesting* with it. This is crucial to their lives as research scientists, because that identity fits them only as long as they are actively *doing* research. All decisions concern possible future action, not mere belief about the past or even the truth about the world in a spectator mode.

NOTES

1. Thanks to Vasso Kindi and Theodore Arabatzis for the invitation to contribute to this volume and to all the colleagues and students with whom I have discussed these issues over the years, most recently Jared Ress and Jonathan Kanzelmeyer in Reno, Dunja Šešelja and Christian Strasser in Ghent, and Octavio Campuzano-Cardona of the Universidad Nacional Autónoma de Mexico, who is working on the role of exemplars in textbooks and in science teaching. Thanks also to the helpful comments by referees. Frank Tobin corrected my Latin. The U.S. National Science Foundation supported my early work on heuristic appraisal.
2. The opening sentence of Shapin's *The Scientific Revolution* (1996) is: "There was no such thing as the scientific revolution, and this is a book about it."
3. While he was critical of then-current accounts of theory reduction and their use in claiming the continuity of scientific change, Kuhn's account of direct modeling may be considered a form of problem reduction but one based at least as much on rhetoric (analogy, similarity, metaphor) as on logico-mathematical derivations. For Kuhn, problem reduction in this broadened sense of "reduction" is more basic than theory reduction (Nickles 2005). Yet the lineages of problems that result provide a kind of continuity through revolution that Kuhn underplays. Or so I shall argue.
4. During the 1960s there was a debate over whether models and analogies in science are purely formal or involve material content as well. Kuhn, Hesse (1966), and Achinstein (1968) rejected the idea, attributed to logical empiricists, that modeling is only a matter of formal, syntactical resemblance. See Nersessian (2008) for a recent defense of material analogy and consider Bird's nice example of an exemplar (Bird 2008, 22–23): a hinged rectangular plate flapping in a flowing fluid. Clearly, a person attempting to solve the problem of the motion of this plate at his or her personal frontier would not first formulate an equation *and then* observe that the equation is symbolically isomorphic to the equation for periodic motion (in this case simple harmonic

motion), as the solution to the problem; for finding that first equation is the very point of the problem. Rather, looking at the diagram (or at the physical process itself, from which the diagram is abstracted), one immediately sees an oscillation or vibration, reminiscent, as Bird notes, of the already familiar motion of a pendulum. The physical analogy is crucial.

5. The term "discovery" has suggested strong realism to some sociologists of science. Neither Kuhn nor I mean it that way any more than sociologists and historians mean to imply strong epistemological realism in talking about the "knowledge" of a scientific community. I employ it as a general term for accepted creative work at the frontier of research. Kuhn himself distinguishes "discoveries , or novelties of fact" from "inventions, or novelties of theory" (52).

6. Hoyningen-Huene (1993), Bird (2000, Chapter 2), and others make valuable points that challenge Kuhn's conservative view of normal science, particularly the tension between Kuhn's treatments of scientific revolutions and unexpected discoveries in normal science.

7. Toulmin also appreciated the virtues of evolutionary models of scientific development. However, I am not committed to his particular conception or to his interpretation of normal science as a quasi-positivist instance of "logicality within a system," except, perhaps, with respect to Kuhn's holistic conception of meaning and conceptual change. This tightly networked conception of meaning made Kuhnian normal science more fragile than need be, with the result that Kuhn had to overemphasize the static nature of normal science. On Kuhn's probable linguistic debt to logical empiricism, see the paper in this volume by Gürol Irzik.

8. Of course, sexual reproduction in biology brings together family lineages. And, because of horizontal gene transfer, biologists today reject a simple tree structure for biological evolution. Another source of messiness in the scientific case is that the modeling relation is not in general transitive, since C may be modeled on B in a different respect than B is modeled on A.

9. Brad Wray (2010, 14) writes that Kuhn's exemplars are flexible. Yes, in the sense that they can be applied to any number of apparently related puzzles, but not in my sense, as far as I can tell. My views have been informed by Vasso Kindi's insightful (2005) and (2010).

10. Kuhn's comparison with legal precedents suggests the same point (7). An attorney reasons that the present case resembles precedent A except in two respects, in which it resembles precedents B and C. But see Kindi (this volume).

11. I cannot here engage the discussion of Max Black's interaction theory of metaphor (Black 1954).

12. See (Nickles 1989, 2006). Ernan McMullin (1976) employs this term. I may have stolen it from him.

13. Hoyningen-Huene (1993, 159, 191–192) writes that such an achievement must satisfy Kuhn's five basic values: accuracy, consistency, scope, simplicity, and fruitfulness. Accepting them as correct problem solutions gives them a local normativity, but it is their estimated fruitfulness, their promise of generating solutions to other problems, that gives them the global normativity that makes them paradigmatic, constitutive of that sort of enterprise.

14. Henderson et al. (2010) attempt to solve the problem of new theories in terms of a hierarchical Bayesian approach that, in effect, reduces heuristic appriasal to retrospective, epistemic appraisal. Interesting and subtle as it is, I believe that their approach involves a sort of regress that fails to solve the problem of new theories and also fails to capture much of what Kuhn had in mind by prospective fertility.

REFERENCES

Achinstein, Peter. 1968. *Concepts of Science.* Baltimore: The Johns Hopkins University Press.

Barnes, Barry. 1982. *T. S. Kuhn and Social Science.* New York: Columbia University Press.

Barnes, Barry. 2003. "Thomas Kuhn and the Problem of Social Order in Science." In Nickles (2003a), 122–141.

Bellone, Enrico. 1980. *A World on Paper: Studies on the Second Scientific Revolution.* Cambridge, MA: MIT Press. (Originally published in 1976, in Italian.)

Bird, Alexander. 2000. *Thomas Kuhn.* Princeton: Princeton University Press.

Bird, Alexander. 2008. "Incommensurability Naturalized." In Soler et al. (2008), 21–40.

Black, Max. 1954. "Metaphor." *Proceedings of the Aristotelian Society* 55: 273–294.

Brush, Stephen. 1988. *The History of Modern Science: A Guide to the Second Scientific Revolution.* Ames: Iowa State University Press.

Ehrenfest, Paul. 1911. "Welche Züge der Lichtquantenhypothese spielen in der Theorie der Wärmestrahlung eine wesentliche Rolle?" Reprinted in Martin Klein (ed.), *Paul Ehrenfest: Collected Scientific Papers.* 1959. Amsterdam: North Holland, 185–212.

Heilbron, John, and Thomas Kuhn. 1969. "The Genesis of the Bohr Atom." *Historical Studies in the Physical Sciences* 1: 211–290.

Henderson, Leah, Noah Goodman, Joshua Tenenbaum, and James Woodward. 2010. "The Structure and Dynamics of Scientific Theories: A Hierarchical Bayesian Perspective." *Philosophy of Science* 77(2): 172–200.

Hesse, Mary. 1966. *Models and Analogies in Science.* Notre Dame, IN: University of Notre Dame Press.

Hoyningen-Huene, Paul. 1993. *Reconstructing Scientific Revolutions: Thomas S. Kuhn's Philosophy of Science.* Chicago: University of Chicago Press.

Kelvin, Lord (William Thompson). 1901. "Nineteenth-Century Clouds over the Dynamical Theory of Heat and Light." *Philosophical Magazine*, Series 6, 2: 1–40.

Kindi, Vasso. 2005. "The Relation of History of Science to Philosophy of Science in the *Structure of Scientific Revolutions* and Kuhn's Later Philosophical Work." *Perspectives on Science* 13: 495–530.

Kindi, Vasso. 2010. "Novelty and Revolution in Art and Science: The Connection between Kuhn and Cavell." *Perspectives on Science* 18: 284–310.

Kindi, Vasso. This volume. "Kuhn's Revolution."

Klein, Martin, Abner Shimony, and Trevor Pinch. 1979. "Paradigm Lost?" *Isis* 70: 429–440.

Kuhn, Thomas. 1960. "Engineering Precedent for the Work of Sadi Carnot." *Archives Internationales d'Histoire des Sciences,* XIII année, nos. 52–53: 251–255.

Kuhn, Thomas. 1962. *The Structure of Scientific Revolutions.* Chicago: University of Chicago Press.

Kuhn, Thomas. 1970a. *The Structure of Scientific Revolutions.* 2nd edition with "Postscript—1969." Chicago: University of Chicago Press.

Kuhn, Thomas. 1970b. "Reflections on My Critics." In Lakatos and Musgrave (1970), 231–278.

Kuhn, Thomas. 1977. *The Essential Tension.* Chicago: University of Chicago Press.

Kuhn, Thomas. 1978. *Black-Body Theory and the Quantum Discontinuity 1894–1912.* Oxford: Oxford University Press.

Lakatos, Imre, and Alan Musgrave, eds. 1970. *Criticism and the Growth of Knowledge.* Cambridge: Cambridge University Press.

Love, Alan. 2009. "Marine Invertebrates, Model Organisms, and the Modern Synthesis: Epistemic Values, Evo-Devo, and Exclusion." *Theory in Biosciences* 128: 19–42.

Masterman, Margaret. 1970. "The Nature of a Paradigm." In Lakatos and Musgrave (1970), 59–90.

McMullin, Ernan. 1976. "The Fertility of Theory and the Unit of Appraisal in Science." In R. S. Cohen, P. Feyerabend, and M. Wartofsky (eds.), *Essays in Memory of Imre Lakatos.* Dordrecht: D. Reidel, 395–432.

Nersessian, Nancy. 2003. "Kuhn, Conceptual Change, and Cognitive Science." In Nickles (2003a), 178–211.

Nersessian, Nancy. 2008. *Creating Scientific Concepts.* Cambridge, MA: MIT Press.

Nickles, Thomas. 1989. "Heuristic Appraisal: A Proposal." *Social Epistemology* 3: 175–188.

Nickles, Thomas. 1997. "A Multi-Pass Conception of Scientific Inquiry." *Danish Handbook of Philosophy* 32: 11–44.

Nickles, Thomas. 2000. "Kuhnian Puzzle Solving and Schema Theory." *Philosophy of Science* 67: S242–255.

Nickles, Thomas, ed. 2003a. *Thomas Kuhn.* Cambridge: Cambridge University Press.

Nickles, Thomas. 2003b. "Normal Science: From Logic to Case-Based and Model-Based Reasoning." In Nickles (2003a), 142–177.

Nickles, Thomas. 2005. "Problem Reduction: Some Thoughts." In R. Festa, A. Aliseda, and J. Peijnenburg (eds.), *Cognitive Structures in Scientific Inquiry: Essays in Debate with Theo Kuipers,* vol. 2. Amsterdam: Rodopi, 107–134.

Nickles, Thomas. 2006. "Heuristic Appraisal: Context of Discovery or Justification?" In J. Schickore and F. Steinle (eds.), *Revisiting Discovery and Justification: Historical and Philosophical Perspectives on the Context Distinction.* Dordrecht: Springer, 159–182.

Reitman, Walter. 1964. "Heuristics, Decision Procedures, Open Constraints, and the Structure of Ill-Defined Problems." In M. W. Shelly and G. L. Bryan (eds.), *Human Judgments and Optimality.* New York: John Wiley, 282–315.

Shapin, Steven. 1996. *The Scientific Revolution.* Chicago: University of Chicago Press.

Sharrock, Wes, and Rupert Read. 2002. *Kuhn: Philosopher of Scientific Revolution.* Cambridge, UK: Polity Press.

Thomson, William, and Peter Guthrie Tait. *Treatise on Natural Philosophy.* Oxford: Oxford University Press, 1867.

Toulmin, Stephen. 1972. *Human Understanding.* Princeton: Princeton University Press.

Wray, K. Brad. 2011. "Kuhn and the Discovery of Paradigms." *Philosophy of the Social Sciences* 41: 380–397.

7 Revolution as Evolution
The Concept of Evolution in Kuhn's Philosophy

Jouni-Matti Kuukkanen

1. INTRODUCTION

In the last chapter of *The Structure of Scientific Revolutions*, Thomas Kuhn suggested that we replace the teleological view of scientific development by an evolutionary image. Kuhn writes that "the developmental process described in this essay has been a process of evolution *from* primitive beginnings. . . . But nothing that has been or will be said makes it a process of evolution *toward* anything" (SSR, 170; emphasis in original).[1] Although the concept of evolution played a role in Kuhn's thinking from the early stages of his career, the later Kuhn took even greater interest in it. The fact that people hadn't ceased viewing science getting closer to something and begun to see it moving away from something troubled him still in the final stages of his life (RSS, 307–308).

Kuhn thought that the "evolutionary turn" in history and philosophy of science would have far-reaching consequences and radically transform our image of science. It would be as decisive a transformation as the one achieved by Charles Darwin in biology with his *Origin of Species* after the mid-nineteenth century (SSR, 171–172). The problem for the early Kuhn was that, although he saw the evolutionary analogy as "nearly perfect," he confesses not being able to "specify in any detail the consequences of this alternative view of scientific advance" (SSR, 171). In this paper, I focus on Kuhn's concept of evolution, trying to find out how far he got in specifying the details. I will be asking how extensively this new concept changes our image of science. Further, I will examine to what degree our view of Kuhn's philosophy is altered, if the concept of evolution is taken seriously. It is likely that all who have been accustomed to viewing Kuhn as a philosopher of radical scientific revolutions will be surprised.[2]

It may be said that a more detailed examination of Kuhn's evolutionary metaphor of science is called for at present. First, the later Kuhn felt that his evolutionary image of science did not get the attention it deserved. In his last interview, Kuhn deplored this situation: "I would now argue very strongly that the Darwinian metaphor at the end of the book [SSR] is right, and should have been taken more seriously than it was" (RSS, 307). This is a very interesting statement, because ten years after the first edition of SSR appeared (1962) there was a relatively active discussion on evolutionary epistemology, which has continued until recent times (e.g., Popper 1968, 1972; Campbell 1974; Toulmin 1972; Hull 1988). What is more, one of his main philosophical adversaries in the 1970s, Popper,[3] was an advocate of evolutionary epistemology. It is likely that Kuhn thought that despite all the existing discussion no one had attempted to develop an adequate evolutionary epistemology in the Kuhnian type of developmental framework. He specifically refers to the missed opportunity to portray scientific *progress* through the evolutionary model; that we should see histories of scientific disciplines, not moving closer to something, but away from their current states, proliferating, multiplying, and specializing (RSS, 307–308).[4]

Kuhn attempted to specify the metaphor a number of times between SSR and the last interview.[5] However, few scholars have examined this aspect of Kuhn's thinking. The only effort to form a more comprehensive picture can be found in Gattei's (2008, 160–163, 168–172) recent book, where he presents a worthwhile overview of what Kuhn said about the concept of evolution.[6] My intention is to further analyze its significance for Kuhn's philosophy of science and for history and philosophy of science in general. Specifically, my aim is to understand how Kuhn's version of evolutionary epistemology relates to or departs from the traditional realism-antirealism-idealism constellation that lurks in the background of this discussion. Third, the significance of this metaphor is only reinforced if one remembers that, when Kuhn discusses his new account of scientific development, he often refers to the book he was preparing.[7] He promised that more details are displayed in it. We know that this unfinished manuscript is entitled The Plurality of Worlds: An Evolutionary Theory of Scientific Development (Gattei 2008, 245; Marcum 2005, 25). In 2010, the book was scheduled to come out soon, under the title *Thomas Kuhn: The Plurality of Worlds*, edited by James Conant and John Haugeland, but its publication was delayed. If Gattei is correct in saying that, in comparison to Kuhn's articles, many of which are in print, there is relatively little new in the book (163), the existing stock of published papers offer an excellent basis for the timely assessment of the meaning and significance of the evolutionary concept of science. On the other hand, the interpretation below may well be deepened and corrected when the book finally appears. In that case, this paper functions as a preliminary step in the path that has still many interesting turns ahead.

In this paper, the first aim is to form an overview of what the evolutionary concept in Kuhn entails. This task highlights two aspects that will be in focus next. The first one is the nature of scientific change. The early Kuhn's wholesale and (psychologically) drastic revolution becomes a gradual and piecemeal communitarian evolution in the later Kuhn, something that may show simultaneous continuity and discontinuity between prerevolutionary and revolutionary stages. The second aspect to be addressed is the concept of incommensurability that retains its central place in the later

Kuhn's philosophy, but receives an uncharacteristically positive role in his theory of scientific development. This is interesting, because incommensurability in science has typically been seen as a threat to scientific progress and rationality. Finally, I attempt to situate Kuhn's evolutionary concept of science in the field of history and philosophy of science. In reference to the evolutionary account of science, Kuhn made some rather striking and radical claims, such as that we abandon the concept of mind-independent world (e.g., RSS, 120, 243). I argue below that we should not view this as idealism; nor should we interpret it as antirealism, although that categorization is possible, because even that conceals some important aspects of Kuhn's evolutionary epistemology. My view is that Kuhn wished us to abandon what might be called a "spectator account of science." In this respect, Kuhn's suggestion resembles the performative notion of science, or the understanding of science as practice, which has become widespread in contemporary historiography of science.

2. REVOLUTION AS SPECIATION

It was already mentioned above that in SSR Kuhn suggested that we replace our image of historical scientific development as teleological by an evolutionary one. According to the younger Kuhn, science does not have a fixed goal, such as truth, toward which it moves, but historical developmental process is a movement away from something. He compared this shift in our worldview to the one achieved by Darwin, who argued that species do not develop according to any preset plan. "*The Origin of Species* recognized no goal set either by God or nature" (SSR, 172).

A good place to continue our examination of Kuhn's concept of evolution is the essay "The Road Since *Structure*." It is future oriented; Kuhn outlines there some of his new ideas, the details of which he promised to specify in his forthcoming book. Yet it also contains commentaries on how his thoughts or interests have changed and developed over the years. Kuhn distinguishes two different types of evolutionary analogy: *diachronic* and *synchronic*. He writes that in his earlier works he was interested in a diachronic parallel between scientific development and Darwinian evolution, that is, "the relation between older and more recent scientific beliefs about the same or overlapping ranges of natural phenomena" (RSS, 97).

The diachronic sense of evolution was designed to highlight two parallels between evolution and the development of science. The first and arguably most important issue for Kuhn was the rejection of teleology in the history of science, that is, the conviction that science is not goal oriented in any sense of the word. Specifically, Kuhn insists that the history of science does not amount to a sequence of improving approximations on the truth (cf. SSR, 170–173; 1984, 244; RSS, 160). The second point that Kuhn emphasized is that the idea of natural selection could be used to illuminate how scientific ideas evolve. In this respect, evolutionary development in the history of science amounts to a view that scientific revolutions are resolved by the selection by conflict of the fittest way to practice science (SSR, 172).[8]

The other, synchronic, parallel between biological and scientific developments "cuts a synchronic slice across the sciences rather than a diachronic slice containing one of them" (RSS, 97). Kuhn is thus interested to describe and compare the relationships between coexisting fields of science. In terms of our image of Kuhn as a philosopher of scientific revolutions, this new focus is very interesting, because the introduction of a synchronic parallel is accompanied by the redefinition of the term "scientific revolution." Kuhn comments on this change by saying that earlier it signified a difference between the cumulative development of science during the normal science and the noncumulative revolutionary science when the old body of accumulated knowledge is overthrown. Now it is a difference between developments, those which do and those which do not require local taxonomic change. Revolution, seen from the synchronic perspective, is speciation of scientific disciplines that results in a proliferation of scientific specialties and in an increase of articulation and specialization. A central feature of speciation is incommensurability between scientific disciplines, a consequence from the emergence of divergent lexicons, which will be in focus in section 3.

Kuhn mentioned speciation and specialization only in passing in SSR (172). In the works of the later Kuhn, speciation and specialization of scientific disciplines come to occupy the center stage. The unit that undergoes speciation is a "community of intercommunicating specialists," who share a lexicon. Speciation means that new fields develop their own lexicons, journals, professional societies, possibly also university chairs, laboratories, and departments. There are two main ways in which a revolution may take place: either a new branch splits off from the parent trunk or a new specialty is formed from an overlap between two preexisting disciplines. This development reveals another parallel with the evolutionary theory: "Over time a diagram of the evolution of scientific fields, specialties, and subspecialties comes to look strikingly like a layman's diagram for a biological evolutionary tree" (RSS, 97–98; similarly RSS, 117, 160).

Kuhn argues that the proliferation of scientific specialties and the emergence of the evolutionary tree of scientific disciplines is an empirical point for which "the evidence, once faced, is overwhelming" (RSS, 250).[9] He mentions (natural) philosophy, medicine, and mathematics as examples of the fields from which various new specialties have split off. According to Kuhn, during late seventeenth-century astronomy, optics, mechanics, geography, and music separate from mathematics. At the same time, speculative chemistry and physics split off from natural philosophy and early biological sciences begin the process of separation from medicine. Kuhn adds that during the nineteenth century these fields acquire the institutional features mentioned above (RSS, 97, 116). Kuhn also uses his personal experience to support his account of speciation. When he left Harvard in 1957, there was only one department of life sciences, but there were already four in 1991. Similarly, when he was studying physics, one physics journal, *The Physical Review*, contained all the major contributions by American physicists and all professionals subscribed to it, but it had split into four by 1991 and few subscribed to more than two journals (RSS, 117). Further, Kuhn mentions two examples of the cases where a new field is born in the intersection of two existing disciplines: physical chemistry and molecular biology (RSS, 97).

The first thing to point out about the relationship between the evolutionary branching of scientific disciplines and the notion of revolution is that revolution as speciation appears to be more restricted in scope than the early Kuhn's concept of revolution. The older sense of revolution included the idea of a widely encompassing transformation, where large bodies of knowledge or entire research programs and orientations are abandoned when scientists jump on the board of a new, more successful paradigm. Second, the old sense of revolution was inherently associated with discontinuity, because the old paradigm is abandoned *in toto* and left in the dust of history, to be possibly returned to only at some unspecified future date.[10]

The earlier Kuhn was confused about whether scientific revolution occurs suddenly and what the unit that undergoes revolution is. On the one hand, he suggested that individual scientists experience a sudden gestalt-switch kind of conversion, after which their world is changed irreducibly. On occasions, he implied that the whole community could go through a similar experience: during revolutions "it is . . . as if the professional community had been suddenly transported to another planet where familiar objects are seen in a different light and are joined by unfamiliar ones as well" (SSR, 111). And yet, Kuhn also recognized that it may take a long time for a revolution to materialize; he says that it took over a hundred years for Copernicanism to become the dominant paradigm (e.g., Kuhn 1957, 227). The older Kuhn wished to set the record straight and admit his earlier mistakes. He pointed out that "gestalt switch" is not a fitting description of scientific revolution, because the entity that undergoes revolution is a community and "gestalt switch" as an individual psychological concept does not apply to sociological entities (e.g., RSS, 48, 88, 242; Kuhn 1993, xiii). Further, in addition to the communal aspect of revolution, the later Kuhn highlighted more clearly that the process toward revolution is gradual in nature. The later Kuhn emphasizes that no lexical changes are "ever vast," but that small changes . . . can have large-scale effects" (RSS, 104; similarly 113–114).

All in all, the new sense of revolution differs from the old one in two specific ways. Unlike the revolutions of the younger Kuhn, there may be continuity between pre- and postrevolutionary periods, and revolutions tend to be localized. First, the consequence of speciation is narrowing of focus, which allows pre- and postrevolutionary continuity. As we saw, speciation may produce an offspring of two preexisting disciplines, which borrows elements from preexisting parent disciplines, or it may result in the formation of an entirely new discipline separated from an older one. A mere narrowing of focus does not suggest total rupture with the previous stage of scientific development. Furthermore, I think this is the context for Kuhn's conclusion that his early attempt to illustrate the evolutionary process by the concept of mutation was wrong. "Mutation" suggests that the emerging alternative is radically different from its ancestor. By contrast, the later Kuhn emphasized that the selection process is adaptionist, where scientific specialties adapt themselves and their tools to their environments. The offspring of adapted species then continue to flourish (RSS, 98; SSR, 149). More important, in interdisciplinary terms, there is continuity, because the old taxonomy or parts of it may remain in use in another discipline (or in some cases maybe in the new one as well).[11] Kuhn refers to a situation where the new split discipline introduces new kinds while the old taxonomy is still used in the old parent disciplines (RSS, 250). Perhaps the clearest example is natural philosophy as an old discipline and physics as a new split specialty in the seventeenth century.[12] Second, speciation also means that only one or two subfields may experience a revolution, while the rest of science and larger research programs are left intact. When a field is born, it does not necessarily cause the transformation of any other fields or disciplines of science, but they may carry on as usual. This is to say that a revolution can be a local phenomenon. The new sense of revolution is thus far from a sudden and comprehensive overhaul of science and its conceptual schemes.

3. FRUITFUL INCOMMENSURABILITY

If we had to choose the topic that was the most important for Kuhn over his whole career, a strong candidate would be the notion of incommensurability. It already appears in SSR, where Kuhn writes that revolutions bring forward scientific theories that are incompatible and often also incommensurable with the prerevolutionary theories (e.g., SSR, 97–103). Incommensurability plays then a central role in Kuhn's theory of scientific speciation, because speciation may lead to taxonomic breaks between disciplines, the relationship of which is characterized by the concept of incommensurability. Indeed, the later Kuhn writes that "no other aspect of *Structure* has concerned me so deeply in the thirty years since the book was written, and I emerge from those years feeling more strongly than ever that incommensurability has to be an essential component of any historical, developmental, or evolutionary view of scientific knowledge" (RSS, 91).

In this section, I will first study how Kuhn understands the notion of incommensurability. After this I consider the significance of this notion. In this respect, the later Kuhn advances a new interesting suggestion. Traditionally, incommensurability has been seen as a threat to the objectivity and progress of science and scientific knowledge, because it has been thought to make theories incomparable. We see that, according to Kuhn, incommensurability is not an impediment to scientific progress but something that enables it. This examination takes us to a consideration of how comparison is possible and what is compared in science. All comes down to a development that creates specialized and separate *incommensurable* disciplines, which continue to advance and progress *in their own niches.*

What Incommensurability Means

Kuhn refers approvingly to Mario Biagioli's paper "The Anthropology of Incommensurability," which outlines the kind of development that leads to incommensurability between two disciplines (RSS, 97). According to Biagioli, members of a discipline try to improve the standing and cohesion of their group after a speciation, which is manifested as unwillingness to learn the language of the other although learning is logically possible by becoming a bilingual. This kind of noncommunicative attitude gives them a better possibility to develop their own language and worldview, which may subsequently lead to incommensurability and "sterility" between the groups (Biagioli 1990, 206–209).

However, it is not clear whether the later (or even the younger) Kuhn's incommensurability is meant to be understood literally. Literally, "incommensurability" means the relation of incommensuration between two or more languages, taxonomies, theories, conceptual schemes, etc., that is, the absence of a common measure. For example, epistemic values, about which Kuhn so often talks, may offer one way to commensuration.[13] Technically, incommensurability for Kuhn means that the no-overlap principle is violated between two taxonomies. The principle says that no terms with different kind labels may overlap in their referents unless they are related as species to genus (RSS, 92; similarly 229–233). As Kuhn says, "there are no dogs that are also cats, no gold rings that are also silver rings, and so on." A language community cannot just add (without lexical revision) a kind term (and its referent) to the old lexicon if it overlaps with the existing referent in the lexicon. The upshot is that the divergence of lexical taxonomies makes translations between them impossible and any unproblematic interlexical communication very hard to achieve. In brief, communication difficulties and the resulting isolation of linguistic communities is the most significant consequence of incommensurability.

The fundamental reason to use the notion of incommensurability in the characterizations of divergent lexicons is that there is no common truth-functional language in which the statements of different lexicons could be expressed and compared. Kuhn claims that some statements may be candidates for being true or false in one lexicon, but not in the other, and that if the lexicons violate the no-overlap principle all the statements of one system cannot be expressed with the language of the other (RSS, 99–100). Kuhn uses the English sentence "the cat sat on the mat" as an example. He claims that it cannot be translated into French, because of the incommensurability between the English and French taxonomies. According to Kuhn, we would need to use several statements in translations; we would need to use such words as "tapis," "paillasson," "carpette," etc., but we could not find one corresponding translation for the sentence (RSS, 93). Kuhn's incommensurability is thus truth-functional untranslatability between divergent lexicons or statements. The English sentence does not have any one translation in French so that it would retain its truth value in all translations. Although explication of different contexts is possible by finding a number of coreferential expressions, no French sentence can express the English one in all contexts due to lexical incommensurability.

Medium of Isolation

What makes the concept of incommensurability a momentous concept in the later Kuhn is its role in the development of science. Kuhn moved to the opposite direction from the mainstream philosophy of science community, where incommensurability is typically seen as a concept that would question science as a rational and progressive endeavor. The idea behind this kind of critical position is roughly the following. If potential rival theories are not commensurable, it is impossible to compare them in any precise manner and choose the better one. Specifically, it is impossible to argue that later theories are better approximations of the truth. Theories simply are not in competition, but merely alternatives to each other in the way that an apple and a banana are. Arguably, one cannot say in any objective manner which is better, and even less, which of them is truer. [14]

Kuhn takes exactly the opposite view of rationality, scientific progress, and incommensurability. "Incommensurability is far from being the threat to rational evaluation of truth claims. . . . Rather, it's badly what is needed, with a developmental perspective, to restore some badly needed bite to the whole notion of cognitive evaluation" (RSS, 91). One may well wonder how this could be the case. In order to see this, we need to examine how Kuhn understands the functional role of incommensurability in the development of science.

In Kuhn, the role of incommensurability in the development of science boils down to its effect in isolating different disciplinary communities from each other. "Lexical diversity and the principled limit it imposed on communication may be the isolating mechanism required for the development of knowledge" (RSS, 98–99). And again, "I am increasingly persuaded that the limited range of possible partners for fruitful intercourse is *the essential precondition for* what is known as *progress* both in biological development and the development of knowledge" (RSS, 99; my emphasis). Kuhn's view is that speciation and the resulting isolation enable scientific disciplines to specialize, which then permits sciences to better focus on some problems and natural phenomena, and develop tools and language for this task. This kind of specialization not only leads to success in individual disciplines, because it "permits the resolution of problems with which the previous structure was unable to deal" (RSS, 250), but it means that science in its totality manages to solve more puzzles than a lexically homogeneous science ever could. "It is by these divisions . . . that knowledge grows" (RSS, 101).

Prerequisite for Progress

One may be surprised to read that there is progress in biological development and in the history of science. Didn't Kuhn deny the possibility of teleological progress in science? It is essential to remember that Kuhn measures progress in terms of puzzles or problems solved. Further, it was mentioned earlier that the evolutionary analogy is designed to explicate the sense in which there is progress in science. Because problems are relative to their disciplines, we should not expect any directional development over several historical periods. In problem solving, the isolation of scientific specialties, a consequence of incommensurability between disciplinary taxonomies, is advantageous. First, it means specialization, and specialization means better tools and better focus on certain problems, resulting in a higher number of problems solved in science. As a consequence of this, it makes good sense to claim that incommensurability is a precondition for scientific progress. It is slightly more difficult to understand how incommensurability could enable, and not disable, rational evaluation in science. In order to see this, let us study the background of Kuhn's claim.

Kuhn positions himself in the philosophical debate on rationality of science by locating himself in the history of philosophy of science. His reading of the history of philosophy of science is the following. Philosophy of science before the historical philosophy of science assumed that rational evaluation presupposes something like an Archimedean platform that can be used to determine the truth or the probability of being true of a particular belief, theory, or law. While Kuhn thinks that there is no Archimedean platform which could be used to assess all theories, he felt that the sociology of knowledge and specifically the "Strong Programme" had "exploited" this conclusion (RSS, 113). Kuhn refuses to accept the sociological explanations of science that rely on the premise "that power and interest are all there are" in scientific development (RSS, 110). In other words, while Kuhn accepts the criticism directed against the internalist explanations of science, prompted by the realization that "observation and experiment were insufficient to bring different individuals to the same decision" (RSS, 108), he maintains that science can be appraised by rational cognitive standards. He is concerned with defending the authority of science as an institution and discipline that uses rational evaluation (cf. RSS, 118–119).

We might say that Kuhn agrees with the critics of incommensurability that meaningful comparison between scientific disciplines is not possible; they are specialized and deal with different phenomena, using divergent taxonomies. However, intradisciplinary comparison and the assessment of cognitively better theories are possible. Kuhn applies the idea of rational comparison to his synchronic evolutionary model that focuses on the cognitive scientific changes of one scientific discipline. His occasional denials of being a relativist and of there being a way to order or rank scientific theories should be seen in this context (e.g., RSS, 160). The idea is that we inquire not after the justification of a belief as such but after a particular change of belief. The comparison is enabled via the set of epistemic values and conducted comparatively between two bodies of knowledge, asking whether the old body of knowledge or the suggested alternative after some changes is better for solving problems (RSS, 96). More specifically, we ask questions like "which of two bodies of belief is *more* accurate, displays *fewer* inconsistencies, has a *wider* range of applications, or achieves these goals with the *simpler* machinery" (RSS, 114).

In the absence of an Archimedean platform and general homogenous scientific taxonomies, Kuhn wishes to relativize rational evaluation to scientific disciplines. He talks about "historically situated Archimedean platforms" (RSS, 95). In order to compare "two bodies of knowledge," all that is required is that they share the same taxonomy or be at least expressible in the same taxonomy or other background belief system. If this can be established, then we can conduct a comparative evaluation of belief changes.[15]

The role of incommensurability is that it isolates linguistic communities and makes them develop homogenous taxonomies that can be used in this task. Incommensurability thus creates the conditions in which scientists can choose a better theory and it therefore enables an *intradisciplinary* progressive development of science. I think this is Kuhn's "middle way" between those who assume that there is one all-encompassing rational standard and those who think that, in the absence of an ahistorical and permanent standard of rationality, social factors determine everything in science.

While Kuhn thinks that specialization and incommensurability as its consequence are something necessary for the development of science, he also admits that it is deplorable if one is trying to find a coherent and unifying worldview. He writes, "Specialization and the narrowing of the range of expertise now look to me like the *necessary price* of increasingly powerful cognitive tools. . . . To anyone who values the unity of knowledge, this aspect of specialization—lexical or taxonomic divergence . . . is a condition to be deplored. But such unity may be in principle an unattainable goal, and its energetic pursuit might well place the growth of knowledge at risk" (RSS, 98; my emphasis). In other words, scientists are forced to introduce new subfields of science. The situation is somewhat paradoxical, because scientists' aim to increase the coherence of their expert fields, which may be achieved by getting rid of the most problematic parts, may result in a decrease in the unity of science. More new fields with mutually incommensurable taxonomies are established as a result (117; see also Kuukkanen 2007).

Finally, the fragmentation of science—"the apparently inexorable . . . growth in the number of distinct human practices and specialties over the course of human history" (RSS, 116)—must be one reason behind Kuhn's proclamations that the development of science is not a process toward anything and that there is no evidence for a process of "zeroing in on nature's real joints" or ontology (RSS, 206). Kuhn is thus saying in effect that, in general, there just is no convergence but unending and nondirectional proliferation and pluralization in basic ontology and concepts in the history of science. Yet as so often with Kuhn, there is more to be said. The examination of Kuhn's antiteleologism in the next section takes us yet deeper into the parallels between scientific and biological developments.

4. SCIENCE AS ADAPTATION

Kuhn promised in SSR that many traditional epistemological problems will vanish, if we adopt the antiteleological evolutionary view of scientific development. It may be an exaggeration to say that they would disappear altogether, but we will soon see that Kuhn was right in the sense that the philosophical landscape looks rather different, if we take his suggestion seriously. In his later texts, he goes on to detail in what way the situation changes. We should keep Kuhn's aspiration to transform the traditional epistemological constellation in our minds, because it helps us to understand what he is getting at.

We have already remarked above that Kuhn's primary reason for rejecting the teleological conception of scientific development is grounded in historical-empirical observations. The empirical history-inspired argumentation was especially prominent in SSR and, more generally, in the work of other historical philosophers at the time (cf. RSS, 95, 111–112). Kuhn argued that the succession of Newton's and Einstein's mechanics does not form a coherent direction of ontological development, and this case was meant as a representative example among many (SSR, 206–207; cf. Kuhn 1957, 264–265; RSS, 111–112). I think Rescher has correctly indicated that the true description of the world should form a maximally coherent whole, a manifestation of ideal coherence (Rescher 1973, Chapter 7; 1985). Specialization or fragmentation of science clearly makes the achievement of ideal coherence or the full true account of nature look like a pipe dream.

However, there is something more behind Kuhn's reasoning. He also thinks that the claims that science is "zeroing in on, getting closer and closer to, the truth" are meaningless (RSS, 243), which may be explained by Kuhn's conviction that "the notion of a match between the ontology of a theory and its 'real' counterpart in nature now seems to me illusive in principle" (SSR, 206). These kinds of statements have triggered the debate over whether Kuhn's fundamental reason for rejecting convergent realism and the correspondence theory of truth was epistemological or metaphysical.

The former suggestion gains some plausibility, because Kuhn is concerned with the evidence of the historical record and problems of access to reality (SSR, 226), and with the problems of how to *measure*, in the absence of an Archimedean platform, whether theories are getting closer to the truth (RSS, 115; cf. Bird 2000, 225–227; Hoyningen-Huene 1993, 263–264; Kuukkanen 2007). However, Šešelja and Straßer have recently suggested that Kuhn's argument against convergent realism was a priori and that Kuhn offered a principled rejection of the correspondence theory. They base their argument on Kuhn's rejection of both the idea of correspondence and the mind-independent world as meaningless (Šešelja and Straßer 2009).

On one occasion, Kuhn explains that the meaninglessness of convergent realism is "a consequence of incommensurability" (RSS, 243–244). The reason is that the divergent lexical taxonomies that we find in the history of science cannot be compared on a shared metric. This expresses the familiar thought that there is no continuity of truth content in the history of science. A new lexicon entails truth statements that are not expressible in another lexicon. Yet, we should say that this position does not make it impossible that, in principle, one or another lexicon is true in the sense of correspondence (whether or not we were able to ascertain this). The real problem is that the history of science contains sequences of alternative ways to structure the world without any apparent cumulation between them.

However, it is true that Kuhn has a still deeper reason to object to convergent realism. Kuhn asks how we should conceive the relation between the lexicon or the shared taxonomy of a speech community and the world. Kuhn argues that we should abandon the notion of an external mind-independent world as it has been employed in the explanations of science in philosophy of science: "no sense can be made of the notion of reality as it has ordinarily functioned in philosophy of science" (RSS, 115). "One big mind-independent world" has to be replaced by something else.

At this point, it is tempting to interpret Kuhn as an idealist, especially if one remembers those famous sentences in SSR on how scientists live in a different world after a paradigm change (e.g., RSS, 111, 117, 121). It is however clear that Kuhn was not an idealist. In fact, he makes it apparent that he does not like the term "mind-dependent world" although he suggested we reject the term "mind-independent world." Kuhn says that the former and the expression "the constructed or invented world" are "deeply misleading" (RSS, 103). "The world is not invented or constructed . . . it is entirely solid: not in the least respectful of an observer's wishes and desires; quite capable of providing decisive evidence against invented hypotheses which fail to match its behavior" (RSS, 101; similarly 120, 159). And it has to be said that this is a reasonable position for a philosopher who takes the evolutionary analogy seriously. No evolutionary biologist would deny that there is a real world with some causal powers although she likely denies that the species reflect some pregiven forms and that they could be evaluated on the basis of how successfully they mirror these "true species."

In my response to Šešelja and Straßer, I suggested that Kuhn found an explanation in the form of historical neo-Kantianism for the phenomena of historical discontinuity and changes of conceptual schemes that the young Kuhn detected in the history of science (Kuukkanen 2009). In other words, the later Kuhn came upon an a priori explanation for his early empirical discoveries. He was attracted to the Kantian idea of categories that "are constitutive of *possible experience* of the world," but he took them in the relativized or historical sense, thinking that they are not absolute and permanent but liable to change over time. Kuhn's structured lexicons are these alternative ways to make the world intelligible.[16] They are the structural products of divergent forms of life within which different propositions may be justifiably taken to be true and false. Further, the historical record of science provides evidence of these various divergent conceptualizations of the world. It is essential that there is no cumulation in transition from one world to another and none of them is truer. Further, the choice between them is pragmatic. (RSS, 245; similarly 248–249)

Kuhn's rejection of realistic discourse is not best seen as a promotion of an antirealist philosophy, and the idealist interpretation is simply wrong. What is at stake is the whole constellation that places scientists on one side and the external world on the other, and assumes that the former try to describe the latter as accurately as possible. This is how "reality" or "mind-independent world" has "ordinarily *functioned* in philosophy of science" (RSS, 115; my emphasis). We might call this traditional view the "spectator account of knowledge,"[17] according to which scientists are like spectators observing what goes on in the object world, from which they are detached. In accordance with the evolutionary theory, Kuhn suggests that we see scientists situated *in* or *within* the world, not in contrast to or opposed to it. The "variety of niches within which the practitioners of these various specialties practice their trade" should take the place of the one big mind-independent world in our explanations of scientific dynamics (RSS, 120).

We should think that each linguistic community creates its own niche. First, the lexical structure of the community structures the world as Kant's categories do. In this sense, Kuhn comes close to the idea that the world is mind-dependent. However, although we need conceptual categories to make the world intelligible, the categories do not make the world. On the other hand, we cannot talk about the unstructured or unconceptualized world. Kuhn's "post-Darwinian Kantianism" is exactly the idea that the lexicon supplies preconditions of possible experience that are liable to change in time and from community to community. This implies the existence of Kant's *Ding an sich*, which really is there although it is "ineffable, undescribable, undiscussible" (RSS, 104). Second, we should see scientists adapting to their environments. Their relationship with the world is not that they observe and describe it from an external standpoint; they live in it and interact with it. They modify their tools of adaptation, but they also modify or carve their niche to fit their needs. "What . . . evolves . . . are creatures and niches together" (RSS, 102; similarly 120). "It is groups and group practices that constitute worlds (and are constituted by them). And the practice-in-the-world of some of those groups *is* science" (RSS, 103). The shared communal lexical structure, which is emulated in individuals' lexicons or mental modules, is one such tool that constitutes the world of the group. Further, lexical structures are assigned a crucial role in the adaptation process. On the one hand, the shared conceptual or taxonomic structure holds the community together. On the other, incommensurability and the resulting communication difficulties have an important function in keeping communities distinct and apart from each other, allowing different groups to specialize, modify, and develop their niche (e.g., RSS, 104, 120).

5. KUHN'S EVOLUTIONARY TURN AND HISTORIOGRAPHY OF SCIENCE

Now we are moving to my final remarks on Kuhn's concept of evolution and I wish to say a few words on the relationship of *Kuhn's evolutionary epistemology* to recent historiography of science. Kuhn's evolutionary turn resembles the movement that has become known as a "practical turn" in historiography of science. We have seen that the crux of Kuhn's argument is that scientific practice and its world are interdefined in the same way as a species and its niche are. In other words, we should not think of the world as independent of the scientific practices that explore it (RSS, 250–251). Kuhn's evolutionary analogy is an attempt to depart from the traditional realist-antirealist discourse, which implies a static object and its observer, in favor of a more pragmatist and situated notion of science, according to which science is practice in a specific environment.

This kind of orientation has become popular in historiography of science since the 1970s. One way to put it is to say that science is taken as a kind of performative action (e.g., Rouse 1987). Another common way to formulate the essentially same view is to say that fundamentally science is a practice that requires *embodied* people, *located* somewhere, *doing* something with some *material* things (e.g., Secord 2004). Sometimes this orientation has been called "constructivism" (e.g., Golinski 1998), which however does not necessarily do justice to this historiographical position, as it does not make one a "social constructivist" in the sense of the term as employed in philosophy of science.[18]

Should we then say that Kuhn was anticipating the practical turn in historiography of science? This would not be correct, because the changes were already underway in historiography of science while Kuhn was publishing. But what we can say is that Kuhn has been one of the most significant influencing factors behind this turn (cf. Rouse 1987; Golinski 1998, especially Chapter 1), sometimes against his own will as he rejected the sociology of scientific knowledge, as we saw above. The difference is that the attention to practice and material things in historiography of science has emerged via historical studies on microphenomena, while Kuhn's emphasis on practice derives from his philosophical thinking. Although a practicing historian may not wish to adopt Kuhn's specific "post-Darwinian Kantianism" with its idea that scientists live in, adapt to, and modify their niches, the latter nevertheless offers one possible theoretical framework for the science-as-practice orientation.

In conclusion, we have seen that Kuhn's evolutionary concept brings many changes into our image of Kuhn. Revolutions that were discontinuous and typically large-scale in the early Kuhn become small-scale speciations that may show continuity. The notion of incommensurability is given a positive, almost a central, role in the development of science. Further, Kuhn's evolutionary epistemology regards scientists as practical achievers, both modifiers and adaptors of their environments, firmly rejecting the spectator account, without falling into idealism. Through the investigation of these topics, I hope to have shown that the later Kuhn's philosophy in general and his concept of evolution in particular develop the themes found in SSR and are resourceful enough to contribute to contemporary philosophy and history of science.

ACKNOWLEDGMENTS

I warmly thank James W. McAllister for his comments on an earlier draft and an anonymous referee for his or her useful suggestions.

NOTES

1. The following abbreviations of Kuhn's books are used: *The Structure of Scientific Revolutions* (SSR), *The Essential Tension* (ET), and *The Road Since Structure* (RSS).
2. See also Nickles (this volume) on a comparison of evolution with revolution in the context of Kuhn's normal science.
3. Popper's *Logic der Forschung* had appeared already in 1934, but was translated into English much later. For the first time it appeared in English as *The Logic of Scientific Discovery* in 1959. In subsequent decades a number of editions have been published.
4. It is worth emphasizing that Kuhn's ideas may be similar to the kinds of theories of evolutionary epistemology that use models and metaphors from evolutionary biology to offer an account of the development of scientific ideas, theories, and disciplines, but not necessarily to the ones that involve a more or less straightforward extension of the biological theory to animal and human cognitive mechanisms (see Bradie and Harms 2008). See also note 6.
5. As far as I have been able to gather, Kuhn discussed the concept of evolution in the following published works: SSR (specifically 170–173) and its 1970 "Postscript" (206); "Reflections on My Critics" (Reprinted in RSS); "Revisiting Planck" (Kuhn 1984, 244–245); "The Trouble with the Historical Philosophy of Science" (reprinted in RSS); "The Road Since *Structure*"

(reprinted in RSS); and "Afterwords" (reprinted in RSS). Many more may, of course, be indirectly relevant. There are other unpublished papers that possibly deal with the notion of evolution, but I have not been able to consult them for this paper (see Gattei 2008, 245, 163n119).

6. James Marcum's book (2005) deserves to be mentioned here, as he also writes about Kuhn's concept of evolution on occasions. Some interesting discussion can also be found in Sharrock and Read (2002, 188–193). Then there are two critical perspectives on the notion of progress in Kuhn's evolutionary model (Bird 2000, 211–214; Renzi 2009). Renzi's paper differs from the rest in that it is an attempt to examine from a biology point of view how well Kuhn's ideas actually tally with the theories and concepts used in evolutionary biology. See Reydon and Hoyningen-Huene's (2010) response to Renzi. I think that they are largely justified to insist that Kuhn meant the concept of evolution as an analogy or a metaphor, not to be taken literally. However, it is not an analogy merely between intellectual processes that took place in the transition from pre-Darwinian to Darwinian biology and the desired transition from convergent progressivism to nonconvergent progressivism on scientific change in philosophy of science (as it may seem in SSR, 170–173). In RSS: Kuhn draws an analogy between different processes in biological evolution and scientific change, although they should not be understood as instantiations of one evolutionary process either. Kuhn is talking on the metaphorical level, in the case of which biological evolution and some of its notions, such as the trajectory of scientific disciplines as a layman's biological tree, scientific communities as isolated developing populations, and scientific development as nonconvergent progress, can be used to illustrate the process of scientific change. Brad Wray's recent book (2011) may offer an attempt to form a comprehensive view of Kuhn's evolutionary epistemology, but it appeared too late to be consulted for this paper. See also his article (2005).

7. For example, essays in RSS: 58, 91, 106, 119, 228, 229.

8. By "fittest" Kuhn refers to the description given in Chapter 12 of *SSR*, where he outlines a process through which revolutions are resolved. The most significant factor in favor of one paradigm is problem-solving ability, but many other play a role as well, such as aesthetic considerations (how "neat" or "simple" a theory is) and future promise (or just "faith" in the theory). Most important, Kuhn presents this process as complex, comparative, and gradually progressing. Persuasive arguments in favor of one paradigm will increase little by little, which leads to a proliferation of experiments, instruments, and articles and books that support it (SSR, Chapter 12, especially 155–159).

9. This claim strikes one as being plausible, when one thinks about all scientific disciplines and subdisciplines that there are currently in science. Rescher interestingly compared how the 1911 and 1974 editions of *Encyclopedia Britannica* described physics. In the eleventh edition in 1911 there are nine constituent branches of physics, which were divided into twenty specialties. The fifteenth edition from 1974 split physics into twelve branches, which had apparently so many subfields that they were not surveyed. Further, The National Science Foundation divided physics in its register of scientific and technical personnel into twelve areas with ninety specialties in 1954, and sixteen areas with 210 specialties in 1970. According to Rescher, substantially the same story can be told for every field of natural science, which is to say that there has been a continual growth in the taxonomic refinement of natural science. Rescher further claims that there is every reason to think that scientific speciation has proceeded exponentially and in the standard evolutionary manner encountered in the biological sphere, just as Kuhn held. (Rescher 1978, 226–230)

10. For example, Kuhn thought that Einstein's mechanics is closer to the Aristotelian than either of them is to the Newtonian (RSS, 206).
11. It is true that Kuhn says that the "new" revolution may also be destructive, like the older concept of revolution implied, when some of the old concepts are replaced by the newer. On one occasion, Kuhn suggests that the old, more comprehensive mode of practice may "die off" (RSS, 120). However, on another occasion, he emphasizes continuity and writes that in another scientific specialty "an evolving form of the old kinds remain in use" (RSS, 250).
12. See RSS (97, 116) and pp. 137–138 above for more examples.
13. One thing is that Kuhn refers to the same set of values repeatedly (e.g., SSR, 152–155, 199 (i.e., "Postscript" of SSR); ET, 322–344; RSS, 114–115, 157–158, 251–252). Epistemic values were thus important for Kuhn since SSR and not just as a way out of incommensurability that his critics had demanded. Indeed, incommensurability gains a central role in the later Kuhn and he defends it as fruitful for scientific progress. Another issue is that he describes them as "necessarily permanent, for abandoning them would be abandoning science together with the knowledge which scientific development brings" (RSS, 252). But see Chang for a different view on epistemic values, in the case of the Chemical Revolution (this volume).
14. The following books are representative examples and summaries of the traditional position on incommensurability: Scheffler (1967); Devitt (1979); Newton-Smith (1981, especially Chapter 7).
15. It is however a good question whether this kind of translatability is actually needed if the epistemic values are interparadigmatic, as Kuhn indicated (see note 13). As long as we are satisfied that two or more theories deal with the same phenomena, and are in competition in this sense, we might attempt to compare them in the absence of common or compatible lexicons. For example, the assessment of the simplicity, consistency, or coherence of a theory does not seem to require the shared vocabularies. One might well be able to choose, say, a more consistent theory even if one is not able to express the theories in one lexical structure. On epistemic values and the coherence theory, see Kuukkanen (2007) and Bonjour (1985, especially Chapter 5).
16. See Andersen (this volume) for a detailed study of the development of Kuhn's account of scientific concepts and conceptual change.
17. The term "spectator theory of knowledge" can be traced to American pragmatism and John Dewey. It is actually surprising that Kuhn did not refer more to pragmatists. Dewey also emphasizes the role of problem solving in human inquiry (e.g., Dewey 1984; this edited volume of Dewey's works contains his 1929 *The Quest for Certainty*). Naturally, the view of science as embodied practice can be seen to connect Kuhn to the later Wittgenstein's philosophy, in which Kuhn was interested. See Read's essay in this volume and his forthcoming book *Wittgenstein among the Sciences* on this issue. Cf. also Read's (Wittgensteinian) interpretation of Kuhn's alleged idealism and Kantianism.
18. Three good overviews of the practical turn in historiography of science are Rouse (1987), Golinski (1998), and Secord (2004). See also Pickering (1992).

REFERENCES

Biagioli, Mario. 1990. "The Anthropology of Incommensurability." *Studies in History and Philosophy of Science* 21: 183–209.
Bird, Alexander. 2000. *Thomas Kuhn*. Chesham: Acumen.

Bonjour, Laurence. 1985. *The Structure of Empirical Knowledge*. Cambridge, MA: Harvard University Press.

Bradie, Michael, and William Harms. 2008. "Evolutionary Epistemology." In Edward N. Zalta (ed.), *The Stanford Encyclopedia of Philosophy*. Winter 2008 ed. http://plato.stanford.edu/archives/win2008/entries/epistemology-evolutionary/.

Campbell, Donald T. 1974. "Evolutionary Epistemology." In P. A. Schilpp (ed.), *The Philosophy of Karl R. Popper*. LaSalle, IL: Open Court, 412–463.

Dewey, James. 1984. *The Later Works, 1925–1953*. Vol. 4: 1929. Carbondale: Southern Illinois University Press.

Devitt, Michael. 1979. "Against Incommensurability." *Australasian Journal of Philosophy* 57: 29–50.

Gattei, Stefano. 2008. *Thomas Kuhn's Linguistic Turns and the Legacy of Logical Empiricism: Incommensurablity, Rationality and the Search for Truth*. Aldershot: Ashgate.

Golinski, Jan. 1998. *Making Natural Knowledge*. Chicago: The University of Chicago Press.

Hoyningen-Huene, Paul. 1993. *Reconstructing Scientific Revolutions. Thomas Kuhn's Philosophy of Science*. Chicago: The University of Chicago Press.

Hull, David. 1988. *Science as a Process: An Evolutionary Account of the Social and Conceptual Development of Science*. Chicago: The University of Chicago Press.

Kuhn, Thomas. 1957. *The Copernican Revolution: Planetary Astronomy in the Development of Western Thought*. Cambridge, MA: Harvard University Press.

Kuhn, Thomas. 1970. *The Structure of Scientific Revolutions*. 2nd enlarged ed. Chicago: University of Chicago Press.

Kuhn, Thomas. 1977. *The Essential Tension: Selected Studies in Scientific Tradition and Change*. Chicago: University of Chicago Press.

Kuhn, Thomas. 1984. "Revisiting Planck." *Historical Studies in the Physical Sciences* 14: 231–252.

Kuhn, Thomas. 1993. "Foreword." In P. Hoyningen-Huene, *Reconstructing Scientific Revolutions: Thomas Kuhn's Philosophy of Science*. Chicago: The University of Chicago Press, xi–xiv.

Kuhn, Thomas. 2000. *The Road Since Structure*. Ed. James Conant and John Haugeland. Chicago: The University of Chicago Press.

Kuukkanen, Jouni-Matti. 2007. "Kuhn, the Correspondence Theory of Truth and Coherentist Epistemology." *Studies in History and Philosophy of Science* 38: 555–566.

Kuukkanen, Jouni-Matti. 2009. "Closing the Door to Cloud-Cuckoo Land: A Reply to Šešelja and Straßer." *Studies in History and Philosophy of Science* 40: 328–331.

Marcum, James A. 2005. *Thomas Kuhn's Revolution. An Historical Philosophy of Science*. London: Continuum.

Newton-Smith, W. H. 1981. *Rationality of Science*. London: Routledge and Kegan Paul.

Pickering, Andrew, ed. 1992. *Science as Practice and Culture*. Chicago: The University of Chicago Press.

Popper, Karl R. 1968. The Logic of Scientific Discovery. New York: Harper.

Popper, Karl R. 1972. Objective Knowledge: An Evolutionary Approach. Oxford: The Clarendon Press.

Renzi, Barbara Gabriella. 2009. "Kuhn's Evolutionary Epistemology and Its Being Undermined by Inadequate Biological Concepts." *Philosophy of Science* 76: 143–159.

Rescher, Nicolas. 1973. *The Coherence Theory of Truth*. Oxford: Oxford University Press.

Rescher, Nicolas. 1985. "Truth as Ideal Coherence." *Review of Metaphysics* 38: 795–806.

Reydon, Thomas A. C., and Paul Hoyningen-Huene. 2010. "Discussion: Kuhn's Evolutionary Analogy in *The Structure of Scientific Revolutions* and *The Road Since Structure*." *Philosophy of Science* 77: 468–476.

Rouse, Joseph. 1987. *Knowledge and Power: Towards a Political Philosophy of Science*. Ithaca: Cornell University Press.

Scheffler, Israel. 1967. *Science and Subjectivity*. Indianapolis: Bobs-Merril.

Secord, James A. 2004. "Knowledge in Transit." *Isis* 95: 654–672.

Šešelja, Dunja, and Christian Straßer. 2009. "Kuhn and Coherentist Epistemology." *Studies in History and Philosophy of Science* 30: 322–327.

Sharrock, Wes, and Rupert Read. 2002. *Kuhn: Philosopher of Scientific Revolution*. Cambridge: Polity Press.

Toulmin, Stephen. 1972. *Human Understanding: The Collective Use and Evolution of Concepts*. Princeton: Princeton University Press.

Wray, K. Brad. 2005. "Does Science Have a Moving Target?" *American Philosophical Quarterly* 4: 47–58.

Wray, K. Brad. 2011. *Kuhn's Evolutionary Social Epistemology*. Cambridge: Cambridge University Press.

8 Incommensurability
Revisiting the Chemical Revolution

Hasok Chang

1. INTRODUCTION

Was there incommensurability between the phlogiston and the oxygen paradigms? The Chemical Revolution of the late eighteenth century was one of the stock examples to which Kuhn referred in *The Structure of Scientific Revolutions* (SSR), although he did not make a separate in-depth study of it. At almost every major juncture in SSR the Chemical Revolution appears as an illustration, including here: "after discovering oxygen Lavoisier worked in a different world" (Kuhn 1970, 118). Therefore the Chemical Revolution has been understood by those sympathetic to the idea of Kuhnian scientific revolutions as a prime case exhibiting incommensurability. Paul Hoyningen-Huene (2008, 101, 114) shows in detail how well the Chemical Revolution fits the Kuhnian model of scientific revolutions, and goes as far as to suggest that this was because the Chemical Revolution was actually constitutive of Kuhn's thinking about revolutions. All the same, in many other accounts of the Chemical Revolution incommensurability does not feature as a key element of the story, and some authors have made an explicit denial of incommensurability in the Chemical Revolution.

So, was there incommensurability in the Chemical Revolution, and if so what exactly did it consist in? In this paper I aim to make a detailed examination of this question. In order to do so, it is first necessary to clarify what is meant by incommensurability. In SSR itself incommensurability was a complex and multifaceted concept, and it is well known that Kuhn himself introduced significant changes to the concept in his later work.[1] The overall conceptual framework I will be using in my analysis rests on a distinction articulated by Hoyningen-Huene and Howard Sankey (2001, ix–xv) between *semantic* and *methodological* incommensurability, which they characterize as follows:

> The thesis of semantic incommensurability derives from the claim of Kuhn and Feyerabend that the meaning of the terms employed by theories varies with theoretical context.

According to the thesis of methodological incommensurability, there are no shared, objective methodological standards of scientific theory appraisal. Standards of theory appraisal vary from one paradigm to another. There are no external or neutral standards which may be employed in the comparative evaluation of competing theories.[2]

My conclusion will be that there was a significant degree of methodological incommensurability in the Chemical Revolution, although only minimal semantic incommensurability.

A study of incommensurability in the Chemical Revolution has threefold significance. For a full historical understanding of the Chemical Revolution itself, incommensurability is a key factor in characterizing the kind of relation that the competing paradigms had with each other. The relation between the paradigms, in turn, was an essential factor shaping the interaction between chemists working in the different paradigms. Second, because the Chemical Revolution has been regarded as an important example of incommensurability, a precise understanding of this case will be instructive and suggestive for general considerations of the nature of incommensurability, and of scientific change more broadly. Third, since the Chemical Revolution, as Hoyningen-Huene points out, was one of the paradigm cases that shaped Kuhn's thinking about scientific revolutions, a better understanding of this case can help us reach a clearer understanding of Kuhn's ideas where it conforms to them, and help us reconsider them where it doesn't.

One fundamental point is worth examining before I attempt to address the main question. Many critics of Kuhn have questioned how broadly applicable the paradigm concept is, and it may be doubted that phlogistonist chemistry or Lavoisierian chemistry really constituted a Kuhnian paradigm. Without getting mired too deeply in the question of what exactly Kuhn meant by "paradigm" in SSR, let us take his own later explication given in *Criticism and the Growth of Knowledge*, namely that a paradigm is either an *exemplar* or a *disciplinary matrix* that grows out of the emulation of exemplars (Kuhn in Lakatos and Musgrave 1970, 271–272). We can identify paradigms in both of those senses well enough in both the phlogistonist and the Lavoisierian traditions of chemistry. On the phlogistonist side, there were exemplars such as Georg Ernst Stahl's work on sulphur (making sulphur by adding phlogiston to vitriolic acid, and recovering the acid by withdrawing phlogiston), and there was an identifiable disciplinary matrix incorporating just the types of elements Kuhn listed—shared symbolic generalizations (e.g., "calcination is the converse of reduction"[3]), shared models (e.g., the metaphysics of principles), shared values (e.g., the commitment to explain the properties of chemical substances), and concrete problem solutions (i.e., exemplars). On the Lavoisierian side, the exemplars included Lavoisier's weight-based analysis of combustion from his "crucial year," and the disciplinary matrix was very well formed partly due to the conscious attempts by Lavoisier and his colleagues to define their new chemistry clearly (consider, e.g., their new definition of the element, their new nomenclature, the distinct chemical roles they assigned to oxygen, their schema of balancing chemical equations by weight, and their predilection for theoretical simplicity and elegance).[4]

2. SEMANTIC COMMENSURABILITY IN THE CHEMICAL REVOLUTION

Taking semantic incommensurability first, let me start by examining a naïve intuition that there was in fact no incommensurability in the Chemical Revolution. Although followers of the oxygen and phlogiston paradigms disagreed with each other seriously, they do not seem to have had much difficulty understanding each other. How was such unproblematic communication possible between competing paradigms? Some have suggested that there was actually a straightforward translation between the languages of the two paradigms. Howard Margolis goes as far as to state: "In general, everything learned about phlogistic processes holds, except that chemists now say that oxygen is gained instead of phlogiston lost, or the converse."[5] Consider combustion (and calcination): where the phlogiston theorists saw phlogiston escaping from the combustible substance, Antoine-Laurent Lavoisier saw oxygen coming in to combine with it. Another example is Lavoisier's theory that oxygen gives acidity to substances, which can be obtained as the mirror image of a fairly common phlogistonist conception that acids are formed by the removal of phlogiston.[6] Shifting between these two viewpoints is akin to a gestalt switch, which was part of Kuhn's initial inspiration for his concept of scientific revolutions; curiously, if all we have is a pure gestalt switch, there is no difficulty with translation. Substituting "phlogiston" with "negative oxygen," as it were, we can straightforwardly translate phlogistonist discourse into oxygenist discourse.

Now, there is something to this naïve commensurability argument, but it does not quite work. For example, Joseph Priestley's "dephlogisticated air" would translate into "oxygenated air" (air with oxygen added to it) rather than "oxygen," which is not quite right. When the phlogistonists say that an acid can attack a metal and disengage phlogiston from it, which creates[7] inflammable air (hydrogen for Lavoisierians), it does not make sense to give a translation to say that one can extract an absence of oxygen, and somehow turn that into hydrogen. Likewise, when the phlogistonists say that various metals have typical metallic properties because they are all rich in phlogiston, it does not quite make sense to render that as an argument that metals are metallic because they all have an absence of oxygen.

A better version of this argument for semantic commensurability can be constructed if we abandon the hope for a full and strict translation, and focus on identifying the common referents of various expressions found in the two theories as and when they occur. This was an idea pioneered by Philip Kitcher (1978, esp. 529–536). Kuhn (1983) criticized this strategy, pointing out that a translation made in that way would do such violence to a text as to make it appear senseless, giving unified treatments of things that had nothing to do with each other. In response Kitcher (1983, 696–697) agreed that through such a translation different tokens of the same expression in the original text would come to be seen as referring to different things. That amounts to a negation of theoretical presuppositions that had united them, and Kitcher regarded disagreements about such presuppositions as the essence of "conceptual incommensurability," which he acknowledged to be widespread but epistemically innocuous. It seems to me that Kuhn broadly followed Kitcher's line of thinking in the end, when he later attempted to define incommensurability in terms of taxonomic structures. From this point of view, incommensurability is not a question of an identity of meaning articulated by literal translation, but a question of whether there is any cross-cutting of boundaries between two taxonomies (or lexicons). Just such cross-cutting of taxonomic boundaries happens, for example, in one of Kuhn's classic cases of incommensurability, namely "planet" in Ptolemaic astronomy and "planet" in Copernican astronomy, as the two concepts do not have the same extension: the set of Copernican planets included the Earth but not the Sun and the Moon, and conversely for the set of Ptolemaic planets.

Chemical taxonomies can also be considered in this way, though substance terms are trickier to handle than object terms. In a study that I find quite convincing, Ursula Klein and Wolfgang Lefèvre (2007) have made a detailed examination of the taxonomy of substances in Lavoisierian and pre-Lavoisierian chemistry, and found that the large categories were preserved intact, although some of them were relabeled and reconceptualized. According to Klein and Lefèvre: "underlying both the phlogistic and anti-phlogistic chemical systems was a shared conceptual structure that remained untouched by Lavoisier's 'revolution'. Thus, in the classification of pure chemical substances that developed in the early modern period, we encounter an ontological deep structure of chemical thinking that was remarkably stable. With respect to this ontological deep structure, no revolution took place at the end of the eighteenth century."[8]

Klein and Lefèvre's examination is thorough as far as it goes, but it is limited in scope because they only focus on substances that are featured in the famous table given by Lavoisier and his colleagues in their work on the new chemical nomenclature. There are some other important concepts to consider. For example, we can identify a class of "phlogisticated" or "phlogiston-rich" substances in phlogistonist chemistry; this class includes a wide variety of substances: charcoal, metals, sulphur, sugar, inflammable air (both modern-day hydrogen gas and carbon monoxide), and phlogisticated air (modern-day nitrogen gas). The question is: are all of these substances classed into one category in Lavoisierian chemistry? And the answer is, actually, yes, though it may not be immediately obvious. Kitcher (1978,

534) points out that in Lavoisierian terms, what these substances all have in common is their strong affinity for oxygen, a suggestion that Kuhn (1983, 674) regards as successful, though with some caveats. All these phlogiston-rich substances react with oxygen and form stable compounds: combustibles burn, metals rust, and nitrogen also makes compounds with oxygen. So, what they have in common is not what they possess in common, but what they *seek* in common (which is, after all, not so far from saying what they lack in common).

In a similar way, we can also consider the taxonomy of *processes* and *operations* involving phlogiston and oxygen, as again suggested by Kitcher (1983, 532) though perhaps not in so many words. The set of operations that constitute adding phlogiston to something coincides with the set of processes that constitute removing oxygen from it. The meanings of "phlogistication" and "removal of oxygen" are very different, but they can both be considered to refer to many of the same processes, including the reduction of metallic calxes, the generation of hydrogen (inflammable air) from water, and the generation of nitrogen (phlogisticated air) from atmospheric air.

Therefore, on the whole, there was a significant degree of semantic commensurability between the two main paradigms involved in the Chemical Revolution, both at ontological and operational levels. But there are many complications and subtleties to note.

(a) Rough edges. First of all, there are some exceptions and qualifications to be noted in the commensurability. For example, while it is true that metals such as iron, lead, copper, and zinc are both phlogiston-rich and rust easily, this is not the case for gold and platinum, which do not rust at all under normal conditions. Gold and platinum are members of the phlogiston-rich set but not members of the oxygen-hungry set. There are also odd cases in the other direction, namely substances that were considered by Lavoisierians to have an affinity for oxygen but not considered to be phlogiston-rich by phlogistonists: for example, Lavoisierians thought that muriatic acid (modern day hydrochloric acid, HCl) had an affinity for oxygen, forming "oxymuriatic acid" (later recognized as the element chlorine, Cl, especially after Humphry Davy's work); phlogistonists generally considered acids to be deficient in phlogiston, if anything. Therefore the phlogiston-content taxonomy and the oxygen-affinity taxonomy did have cross-cutting boundaries. It is tempting to dismiss the case of muriatic acid as one in which the Lavoisierians simply had a misconception of what happens when it loses hydrogen, but gold is not a very comfortable exception to have here, as it was considered the most perfect of all metals. Also, Priestley's "phlogisticated air" (modern-day nitrogen gas) is clearly phlogiston-rich, but it will not combine with oxygen gas if the gases are left to their own devices, so its oxygen affinity is questionable (although elemental nitrogen and oxygen will combine with each other to make stable compounds). Actually most combustible substances will not combine spontaneously with oxygen; they need to be set on fire or rubbed strongly or something—that is to say, they only have strong

affinity with oxygen at high temperatures, although high phlogiston content is not a temperature-dependent property. On the whole, the correspondence between being phlogiston-rich and oxygen-hungry is not exact or universal. I think the best way of viewing this situation is to apply the notion of "graded structures" as employed by Hanne Andersen, Peter Barker, and Xiang Chen (2006, 8ff.): "human subjects readily rate instances of a given concept as better or worse examples of the concept." Without pretending to do justice to the subtleties of Andersen, Barker, and Chen's treatment, I would like to note that their use of graded structures allows them to take incommensurability as a matter of degrees and something that can occur at various levels in taxonomic hierarchies (2006, Chapter 5).

(b) Multiple commensurabilities. If the only commensurability relation we can discern is an imprecise or imperfect one as noted above, then commensurability is a rather promiscuous relation, and not even strictly transitive. One paradigm may be commensurable with multiple other paradigms, which may not even be commensurable with each other. In the later phases of the Chemical Revolution, this was perhaps more than an abstract in-principle possibility, as there were many articulations of the phlogiston paradigm, and it is probably the case that all of them had a fairly good degree of semantic commensurability with the oxygen paradigm. (On the oxygen side there was much more unity due to Lavoisier's strong doctrinal control.) So far I have considered only one version of the phlogiston theory, which Alan Musgrave attributes to Henry Cavendish's work published in 1766; Cavendish himself modified his theory by 1784, and Priestley followed Cavendish in this modification.[9] The revised theory remained semantically commensurable with Lavoisier's theory. With Cavendish's 1766 version of the phlogiston theory, the Lavoisierian operation of oxidization corresponded to the phlogistonist operation of phlogiston removal. With the 1784 version, phlogiston removal was reconceptualized as a more complex operation of removing phlogiston and adding water (which is equivalent to adding dephlogisticated water), but this, too, had a good correspondence with Lavoisierian oxidization.

Interestingly, we can even see good commensurability between the phlogiston chemistry and post-Lavoisierian modern chemistry. The English chemist William Odling (1871) saw a good correspondence between phlogiston and chemical potential energy.[10] Combustibility corresponds to having potential energy that can be released at a high intensity, with or without the help of oxygen. This way of thinking is quite commensurable with the phlogiston theory, but not with Lavoisierian theory. Another point of commensurability was noted by the American chemist G. N. Lewis (1926, 167–168): phlogiston corresponds to electrons, or rather, phlogistication corresponds to the gaining of electrons, and oxidation corresponds to the losing of electrons. After all, "oxidation" in modern chemistry is a misnomer, since it does not necessarily have anything to do with oxygen. If we draw a parallel between oxidation/reduction and dephlogistication/phlogistication, there is very good commensurability between the phlogiston theory and the modern chemistry of redox reactions, actually much more than between modern chemistry and Lavoisierian chemistry.[11]

(c) Sense vs. reference. As recognized clearly by Kitcher, Kuhn, and others, the kind of commensurability discussed here only gives us correspondence in reference, not a literal translation of sense-meaning. Therefore the designation of this as "semantic" commensurability may be misleading, for those to whom "semantic" indicates something to do with "meaning" in its intuitive sense, including its nonextensional dimensions. To say that two theories have commensurable lexicons only means that they have descriptions that refer to the same sets of objects and processes, not that the object descriptions and process descriptions have entirely the same meaning.

This gap between sense and reference is what allows us to get around the difficulty of the naïve commensurability argument. For example, although we cannot say that the inflammable air produced from reacting a metal with an acid somehow consists of the absence of oxygen, it does make sense to say that what the phlogistonists saw as a process of phlogiston removal (from the metal) coincides with what the Lavoisierians saw as a process of the metal combining with oxygen (the inflammable air being the hydrogen left behind when oxygen from the water in the acid is taken away by the metal). This also makes sense of how various chemists could see combustion as a dual process involving *both* the loss of phlogiston and the combination with oxygen; Whiggishly speaking, we have to say that these latter chemists had the best description of the process, although all sides referred to the same process in nature.[12]

(d) Interpretation. What points (b) and (c) also indicate is that the establishment of commensurability in these situations requires interpretation. We saw that an attempt to draw a simple compositional equivalence between phlogiston and anything in the Lavoisierian system fails. But we can *establish* commensurability by finding the right interpretation, that being rich in phlogiston is equivalent to having a strong affinity for oxygen, or that subtracting phlogiston from something corresponds to adding oxygen to it (or adding water and subtracting hydrogen). These interpretations are not fully determined by the simple extensional relations involved; rather, it is the interpretations that frame the extensional relations. This is quite in line with Kitcher's call for us to "abandon the search for a CIT [context-insensitive theory of reference]" and embrace "context-sensitive theories (CST's)" (Kitcher 1978, 535, 523–525). Kuhn (1983, 674) made the objection that such context-dependent reference determinations did not deserve to be called "translations," but that is neither here nor there if (in)commensurability is not only about translations as such, as Kuhn's original formulation in SSR allowed (see the discussion of methodological incommensurability in section 3 below).

(e) Levels of description. Thinking further about the sense–reference distinction leads to an important qualification in the commensurability

argument. What we have to acknowledge is that there are multiple possible levels of sense-meaning, and two paradigms may be quite commensurable at one level but not at another. In the Chemical Revolution and many other instances of scientific discourse, I can distinguish at least three levels of description (though they often occur in a mixture), which I will call operational, phenomenal, and theoretical. At the operational level we have descriptions of the procedures that we perform: mix up some ingredients, notice the air that bubbles up, collect it in an inverted beer glass, put a mouse in it to see how long it lives. At the phenomenal level we have descriptions of our objects of study, based on their observable properties: so we have inflammable air, metals and calxes, etc. At the theoretical level we have descriptions of processes and objects couched in theoretical terms: inflammable air is really phlogisticated water, or hydrogen (water maker); etc. The theoretical level also includes descriptions of unobservable processes and objects that are presumed to exist; some phlogistonists believed that all the different subtle fluids were various modifications of elementary fire; on the basis of his theory that all acids contained oxygen, Lavoisier ([1789] 1965, 175) proudly put the "muriatic radical" (hydrochloric acid minus oxygen) in his table of elementary substances.

What I have argued so far is that there was good commensurability between the competing paradigms in the Chemical Revolution, at the operational and phenomenal levels. But at the theoretical level of description, commensurability failed clearly. If we take Lavoisier's muriatic radical, there was no operational method of finding its reference class so that one could see how the phlogiston theory would describe members of that reference class. Only a theoretical description of its extension was available, namely "what one gets when one takes oxygen away from muriatic acid," and that theoretical description does not correspond to anything definite in the phlogiston theory. As Kuhn (1983, 675–676) points out, there is a similar lack of commensurability for some very general theoretical categories of phlogistonist chemistry, such as "elements" and "principles," which had no clear equivalents in the oxygenist paradigm.[13]

(f) Commensurability needed for incommensurability. Thinking about levels of description reveals an interesting general feature of Kuhn's later notion of semantic incommensurability. When we say that incommensurability is defined by whether reference classes have cross-cutting boundaries, there is a presumption that the members of those reference classes can be identified unambiguously. In other words: if the question of commensurability is "does paradigm *A* classify entities into the same groups as does paradigm *B*?" we can answer (or even ask) that question only because we are agreed on how to identify these entities (like the individual birds in Kuhn's ducks/geese/swans illustration). So it seems that the question of incommensurability at one level of description can be asked and answered only if there is commensurability in this sense at a lower level of description.

In the case of the Chemical Revolution, the commensurability was quite strong at the phenomenal level, and rock-solid at the operational level. Those levels provide a good platform from which we can ask about whether there was incommensurability at the theoretical level. But this also implies that the perceptual aspect of incommensurability ("world changes") was quite minimal in the Chemical Revolution. There was a great deal of continuity at the operational and phenomenal level through the Chemical Revolution, and this is also one important reason why the various protagonists generally had no difficulty in understanding each other. Basic communication was easily possible by going down to the phenomenal level, and to the operational level if necessary.

Where do all these elaborations leave us? To conclude, I would say that there was a good deal of semantic commensurability in the Chemical Revolution, though it was not perfect except possibly at the operational and phenomenal levels. Does this mean that Kuhn was wrong to emphasize incommensurability in the Chemical Revolution? I do not think so, because I think there was strong and significant methodological incommensurability.

3. METHODOLOGICAL INCOMMENSURABILITY IN THE CHEMICAL REVOLUTION

Methodological incommensurability is about standards of evaluation, and there are multiple dimensions to it, as indicated in Kuhn's writings, most of all SSR. I will take these dimensions one by one.

3.1. Problem Field

In SSR it was one of Kuhn's main points regarding incommensurability that different paradigms have different lists of problems they consider legitimate and important.[14] Considerations of problem field also cover much of what is often referred to as "explanatory power." This most indisputable element of incommensurability often got swept aside in the excited discussions about other aspects.[15] There will be some problems that are commonly acknowledged to be important in all paradigms, and others that are considered important only by some but not others. Table 8.1 below shows the division of the problem field in the Chemical Revolution.

Both the oxygen paradigm and the phlogiston paradigm assigned great importance to the understanding of three processes recognized by both sides as closely related to each other: combustion, calcination, and respiration. The theory of acids was also something considered important by both sides, though perhaps particularly important to Lavoisier. Richard Kirwan (1789, 38), in his classic exposition of the phlogiston theory, considered that Lavoisier's work on acidity was an important contribution: "With respect to the nature and internal composition of acids, it must be owned that the theory of chymistry has been much advanced by the deductions and reasonings of Mr. Lavoisier." Closely related to these issues was the constitution of various substances, including water, metals and calxes, various kinds of "airs," and also various nonmetallic substances that had been considered to be phlogiston-rich. Kirwan (1789, 6–7) thought that constitution was the key to the whole dispute: "The controversy is therefore at present confined to a few points, namely, whether the *inflammable principle* [phlogiston] be found in what are called phlogisticated acids, vegetable acids, fixed air, sulphur, phosphorus, sugar, charcoal, and metals." Similarly, in his latter-day defense of the phlogiston theory published from his exile in Pennsylvania, Priestley ([1796] 1969) focused almost exclusively on constitution—of metals and water most of all, but also of carbon, nitrogen, and fixed air. In these works Kirwan and Priestley were responding to Lavoisier's contention that he had refuted the phlogistonist conceptions on the constitution of these substances.

In contrast to the above problems, some other problems were not universally considered important. The phlogiston paradigm shared a main preoccupation of pre-Lavoisier chemistry, which was "to account for the qualities of chemical substances and for the changes these qualities underwent during chemical reactions" (Kuhn 1970, 107). More specifically, phlogistonists sought to explain the properties of compounds in terms of the "principles" that entered into their composition. In the oxygenist paradigm this was not regarded as a legitimate question, and chemistry reclaimed this stretch of territory only in the twentieth century. One salient case was the explanation of why metals (which were compounds for phlogistonists) had a set of common properties (Kuhn 1970, 148). Actually by the onset of the Chemical Revolution this was no longer a research problem in the phlogiston paradigm, as it was accepted almost as common sense that metals had their common metallic properties (including shininess, malleability, ductility, and electrical conductivity) because of the phlogiston they contained. The

Table 8.1 The Division of the Problem Field in the Chemical Revolution

Problems considered important by both sides	Problems considered important only by phlogistonists	Problems considered (very) important only by oxygenists
· Understanding of combustion, calcination/ reduction, and respiration · Theory of acids · Constitution of various substances	· Explaining properties of compounds in terms of properties of ingredients · Nature and common properties of metals · Mineralogy; geology · Meteorology · Nutrition; ecology	· Theory of heat and changes of state · Chemistry of salts

oxygenist side seems to have rejected not so much this answer as the problem itself. As Hoyningen-Huene puts it (2008, 110): "only after more than a hundred years could the explanatory potential of the phlogiston theory be regained in modern chemistry." There were also some other problems considered important by *some* phlogistonists, although not all phlogistonists were equally concerned about them. These included various problems in mineralogy, geology, and meteorology.

On the other side, there were also some problems considered much more important by oxygenists than by phlogistonists. Thermal phenomena were noted by all chemists, and various phlogistonists tried to give some account of the nature of heat, but it was Lavoisier, building on Joseph Black's work on latent heat, who really brought heat (caloric) centrally into chemistry. Closely related to the theory of heat were questions regarding what we now call changes of state, about which Lavoisier had a very definite theory, to which he gave a very prominent place in his system of chemistry.[16] A similar case is the chemistry of salts. This was a common preoccupation of eighteenth-century chemistry, but the phlogiston theory had relatively little to offer in this area. In contrast, this area of research held out much promise for the new chemistry on the basis of the apparent triumph of Lavoisier's theory of acids, as one can tell from the fact that Lavoisier devoted an entire third of his *Elements of Chemistry* to it, under the heading "of the combinations of acids with salifiable bases, and of the formation of neutral salts."[17]

How can we compare the performance of the two sides in this fragmented problem field? Rather predictably, and just as Kuhn said, each side tended to provide good solutions to the problems that it considered important, and not such good solutions to other problems. Doing well on problems that the other side does not recognize as important does not have much persuasive force, but it does strengthen the belief in one's own side. Therefore it tends to prolong the dispute, rather than resolve it. On the other hand, if one side does clearly better than the other in addressing the commonly recognized problems, that constitutes a clear advantage in the competition as judged by the actors themselves. In the Chemical Revolution, there was no shortage of common problems but each side thought it was doing quite well on all of them, at least as well as the other side. This divergence of assessment was not caused by any semantic incommensurability that interfered with the comparison of solutions between paradigms. On the contrary, semantic commensurability helped prolong the debate in cases when translations seemed to reveal that there was no ultimate difference between what the two paradigms were saying. For instance, Cavendish (1784) stated that he saw no reason to switch over to the Lavoisierian paradigm since he could make perfect sense of what the latter said in cases like the composition of water by translating "oxygen" as "dephlogisticated water" and "hydrogen" as "phlogisticated water." In other cases the two paradigms did offer substantively different solutions to common problems, but disagreed on the evaluation of the relative qualities of those solutions. This brings us to another key dimension of methodological incommensurability, concerning the criteria of judgment.

3.2. Epistemic Values

Turning now to the assessment of the quality of problem solutions, we can easily see that there were divergent epistemic values involved. The most important factor was the opposition between *simplicity* and *completeness*. Oxygenists, particularly Lavoisier himself, prized simplicity greatly—especially the kind of simplicity that one could describe as elegance. Phlogistonists, particularly Priestley, saw more importance in completeness, wanting to account for all the observed phenomena in a given problem area and for all the observed aspects of those phenomena. Lavoisierians liked to focus their attention on paradigmatic cases in which their theoretical conceptions worked out beautifully, leaving the messier cases aside. In contrast, Priestley and some of his fellow phlogistonists endeavored to explain all the major phenomena they produced and observed, even if the explanations got cumbersome in the more difficult cases.

A good example to illustrate this point is the calcination and reduction of metals. The red calx of mercury, which Priestley had initially used when he produced dephlogisticated air, was taken up by Lavoisier and his colleagues as the paradigm case showing that calcination and reduction were processes of oxidization and deoxidization. Mercury could be turned into this red calx by heating in ordinary air; the calx could then be turned back into metal simply by a higher degree of heat (produced by a large burning lens), yielding oxygen and producing or absorbing no other detectable substances. This wonderful exhibition of oxidization and reduction was cited over and over by the Lavoisierians.

Priestley protested ([1796] 1969, 24): "But this is the case of only this particular calx of this metal." In his view, the Lavoisierians were distorting the whole picture by focusing on one exceptional case. Other metals behaved differently; Priestley (31) pointed out that no calx of iron could be revived "unless it be heated in inflammable air, which it eagerly imbibes, or in contact with some other substance which has been supposed to contain phlogiston." Even for mercury, there was another type of calx[18] "which cannot be completely revived by any degree of heat, but may be revived in inflammable air, which it imbibes, or when mixed with charcoal, iron-filings, or other substances supposed to contain phlogiston" (24).

Lavoisier, for his part, could not tolerate the continual complications and mutually conflicting changes that various phlogistonists introduced into their theories in their attempts to meet the challenges posed by various new phenomena:

> Why, therefore, need we have recourse to an hypothetical principle, the
> existence of which is ever supposed, and has never been proved; which in

one case must be considered as heavy, and in another as void of weight, and to which, in some cases, it is necessary even to suppose a negative weight; a substance which in some instances passes through the vessels, and in others is retained by them; a being which its maintainers dare not rigorously define, because its merit and its convenience consist even in the uncertainty of the definitions which are given of it?[19]

It is not that either side failed to recognize the desirability of either simplicity or completeness, but there were clear differences in the degree of emphasis, or preoccupation, with those competing values. This divergence between simplicity and completeness also played an important role in the debates regarding combustion. There were very serious anomalies in Lavoisier's theory of combustion, as I have discussed in detail in another paper,[20] but Lavoisierians chose to brush aside the anomalies instead of complicating their theory in order to accommodate them.

There were also broader types of epistemic values at play. A sort of epistemic conservatism was one of the values upheld by many phlogistonists, while oxygenists were taken with the idea of reform or novelty in itself. There is an interesting passage in Cavendish illustrating this point (1784, 152): "it will be very difficult to determine by experiment which of these opinions is the truest; but as the commonly received principle of phlogiston explains all phenomena, at least as well as Mr. Lavoisier's, I have adhered to that."[21] Cavendish's temperament shown here is surely an informative contrast to the youthful enthusiasm of Lavoisier, declaring to himself in 1773 that his investigations were "destined to bring about a revolution in physics and chemistry," before he had published even his first attack on phlogiston.[22]

On the other hand, it cannot be that the leading phlogistonists were simply opposed to scientific change, as they surely delighted in making new discoveries and crafting some new theoretical ideas, too. I think it is fair to say that many arguments made on behalf of phlogiston were actually motivated by *pluralism*, in reaction against Lavoisierian dogmatism. This is quite contrary to the common notion that the phlogistonists were blinded by dogma. Alfred Nordmann (1986) has drawn attention to Georg Christoph Lichtenberg's antidogmatic protest against the Lavoisierians' attempt to close the debate prematurely. The pluralist tendency was even stronger in Priestley, and his 1796 defense of phlogiston is a moving testimony in this regard. Declaring that "free discussion must always be favourable to the cause of truth," he reminded the reader of the nondogmatic path he had walked in science:

No person acquainted with my philosophical publications can say that I appear to have been particularly attached to any hypothesis, as I have frequently avowed a change of opinion, and have more than once expressed an inclination for the new theory, especially that very important part of it *the decomposition of water.* (Priestley [1796] 1969, 21)

He dedicated his book to "the surviving answerers of Mr. Kirwan" (the post-Lavoisier leaders of French chemistry, namely Berthollet, Laplace, Monge, Guyton de Morveau, Fourcroy, and Hassenfratz), and requested an answer from them to his objections to the oxygen system. Priestley drew an ominous parallel between the politics of science and the larger politics that had put a premature end to Lavoisier's life: "As you would not . . . have your reign to resemble that of *Robespierre*, few as we are who remain disaffected, we hope you had rather gain us by persuasion, than silence us by power."[23]

I do not think that all this was retrospective self-fashioning or a loser's spiteful plea for survival. Priestley had expressed similar epistemic views even at the height of his fame and success. For example, in his letter of May 25, 1775, to Sir John Pringle, the President of the Royal Society, in which he announced the discovery of dephlogisticated air (oxygen), Priestley wrote (1775, 389):

> It is happy, when with a fertility of invention sufficient to raise *hypotheses*, a person is not apt to acquire too great attachment to them. By this means they lead to the discovery of new facts, and from a sufficient number of these the true theory of nature will easily result.

This passage directly followed his proposal that "nitrous acid is the basis of common air, and that nitre is formed by a decomposition of the atmosphere," to which he added: "But I may think otherwise to-morrow." One can almost hear an echo of Montaigne finishing his thoughts with "though I don't know."[24] In contrast, there was a clear absolutist impulse on the oxygenist side, perhaps most egregiously manifested in the ceremonial burning of Stahl's phlogistonist text.

Before leaving the issue of epistemic values, I should note that even one and the same epistemic value may be interpreted and instantiated in divergent ways, leading to very different conclusions and even mutual accusations of betrayal of the value in question, as Kuhn himself emphasized (Kuhn 1977, 331). There are a few significant cases of divergent value-instantiation in the Chemical Revolution. First of all, both sides valued unity, and each side cited the kind of unity it was able to achieve as persuasive evidence in its own favor. There was some convergence in this, since both theories united combustion, calcination, and respiration in similar ways. But beyond that there was significant divergence in what was unified, and how. Lavoisier's caloric theory united combustion and changes of state. His ideas on oxygen united combustion and acidity, since many combustion products were acidic. On the phlogiston side, there was a pleasing theoretical unity concerning the behavior of metals: their common properties, their calcination/reduction, and their reaction with acids. The phlogiston theory was also much more conducive to a grand unity of all imponderable substances, as manifestations of "elementary fire": phlogiston, electricity, light, heat, magnetism, and so on.

In a similar vein, the adherence of both sides to the value of systematicity is manifested in each side accusing the other of arbitrariness. Lavoisier's disdain for what he regarded as arbitrary shifts in phlogiston theory is apparent. On the phlogistonist side, the accusation against the Lavoisierians was that they failed to adhere to the rule of assigning like causes to like effects. Both Priestley ([1796] 1969, 33) and Kirwan (1789, 281–282) used this argument in constitutional debates, to combat the oxygenist refusal to recognize the common presence of phlogiston in various substances.

An examination of the arguments on both sides also reveals their common allegiance to what I will call "empiricism": using no extraneous hypotheses, just empirical facts and ideas derived from facts. In a passage already quoted, Lavoisier and his colleagues denounced phlogiston as a hypothetical entity whose existence was "ever supposed, never proved." In their own theory, "nothing is admitted but established truths"; Lavoisier's was "a doctrine which explains all the facts of chemistry without any supposition."[25] It is difficult to take the latter claim seriously, as there were blatantly hypothetical entities such as caloric in their own system, but that does not negate the place of empiricism within the Lavoisierian ideology. Most phlogistonists were no less adamant about their empiricism, and perhaps practiced it more realistically and with less self-righteousness. I have already quoted Priestley as saying that he did not have strong attachments to hypotheses, and regarded them mostly as means for eliciting new facts.

3.3. Practice-Based Metaphysics: Principlism vs. Compositionism

There is one crucial aspect of the Chemical Revolution that I have not dealt with yet. And the consideration of this factor will bring us to an entirely different dimension of incommensurability. Many people have argued that Lavoisier's most important contribution to chemistry was his emphasis on weights: his recognition of its importance, and his use of precision measurements to trace it through chemical changes. For Lavoisier and his followers, the most decisive arguments against the phlogiston paradigm were based on the considerations of weights in chemical reactions. In contrast, it seems that the phlogistonists were not so centrally concerned with weight, though some of them did recognize it as a relevant property and even knew how to measure it extremely well.[26] The difference here is not quite a matter of general epistemic values, but something more specifically tied to certain types of scientific practice.

My view is that Lavoisier's emphasis on the chemical arithmetic of weights originated in, and derived strength from, a type of experimental–theoretical chemical practice which I call "compositionism," which is to be contrasted to the "principlism"[27] practiced by many phlogistonists.[28] Principlism held that there were *principles*, namely fundamental substances that impart certain characteristic properties to other substances; in that ontology, there is an asymmetry between principles and the other substances that are transformed by them, principles being active and the others passive.[29] Experiments framed in principlist thinking treated chemistry as a business of *transformations*: this was partly rooted in the alchemical notion of the transmutation of substances from one to another, but the principlist tradition had more practical roots, too, as we will see in a moment. In contrast, compositionism was based on two key ideas: the existence of stable and enduring chemical substances, and an equal ontological status of all basic chemical substances. The main compositionist idea was that chemical reactions were processes of taking substances apart and rearranging the parts to make up other substances, or putting the parts back together to recompose the original substance.

For the compositionists, the *constitution* of a chemical substance is reduced to the question of *composition*. When different chemical parts come together to make a whole, they do not interact so as to change each other's natures; they somehow remain intact in the compound that they make up, and can be separated out from each other, having suffered no changes. It could be said that the compositionist tradition had more respect for nature than did the principlist tradition, at least compared to the alchemical strand of the latter; in the compositionist tradition, there was no conceit about human "art" being able to make anything truly new. So, for instance, compositionist pharmacy concentrated on *extracting* the medicinally valuable components from naturally occurring materials. Compositionist mineralogy and metallurgy focused on extracting metals from ores and separating different metals from each other, rather than transforming one metallic substance into another.

We can have a more nuanced view of the contrasting fortunes of the oxygenist and phlogistonist paradigms, if we keep in mind the general opposition between compositionism and principlism, and the slow yet sure rise of compositionism through the eighteenth century. One can have a stark view about which side was right in the Chemical Revolution if one shares Lavoisier's compositionist preoccupation with weight and his conception of weight as a conserved quantity. Take the decomposition and recomposition of water. Lavoisier's view is crystal clear: we take 100g of water, and we make 15g of hydrogen and 85g of oxygen out of it;[30] and then, we can put those precise amounts of hydrogen and oxygen together, and make 100g of water again. What better proof than this could one have for the idea that water is a compound made up of hydrogen and oxygen? Compare this beautiful account with the phlogistonist story: hydrogen is meant to be "phlogisticated water" and oxygen "dephlogisticated water." Cavendish and Priestley thought that water was the base of all gases, but gave no convincing story about why phlogistication should make water less dense than dephlogistication does. There is no definite account of how much phlogiston needs to be given to a given amount of water to make hydrogen, and how much needs to be given in order to make oxygen. One can easily see the force of Lavoisier's weight-based account.

We must understand how the likes of Priestley were able to resist the above way of thinking and maintain that weight was not the most important parameter in chemical reactions, if we are to get to the bottom of the methodological incommensurability in the Chemical Revolution. Lavoisier's reasoning rested on two very significant assumptions:

(a) Weight is a good and proper measure of the amount of all chemical substances.
(b) Weight is conserved.

But most phlogistonists would not have accepted assumption (a), since phlogiston was an imponderable (weightless) substance. Ironically, assumption (a) is not even entirely appropriate for Lavoisier's own chemistry, as the first two in his list of simple substances, light and caloric, did not have weight. Assumption (b) is also tricky. Contrary to common intuitions, there never was any inexorable reason why weight must be conserved (and indeed, $E=mc^2$ proclaims that it is not). Priestley and some others were not ready to accept the precise weight conservation claimed by Lavoisier in the decomposition and recomposition of water. For those who believe today's chemistry, it should be impossible not to sympathize with Lavoisier's critics, as we witness Lavoisier's utter conviction that 85:15 was the correct oxygen–hydrogen ratio in water, rather than anything like 8:1. The only thing I can conclude is that assumptions (a) and (b) were *presumed* as part of compositionism, not supported directly by their own evidence. Was compositionism on the whole supported by sufficient evidence? Not in a simple way. Principlist and compositionist ways of thinking were linked up with corresponding experimental practices, many of which were not directly related to the dispute between the phlogiston and the oxygen paradigms, and extended far beyond the scope of the oxygen–phlogiston dispute. When I say "linked," what I mean is that the conceptual and the experimental aspects reinforced and shaped each other, rather than one causing the other in a unidirectional way.[31]

Here I will just give a brief initial view of the joined-up experimental–metaphysical systems underlying the phlogiston and oxygen paradigms. For the phlogistonist side of this story, we can look at Priestley again. His pneumatic chemistry was a transformative experimental practice. Before he started making the discoveries of all the various different "factitious airs" that made him famous, he was making observations on things like "air *infected* with animal respiration" and "air *infected* with the fumes of burning charcoal" (Priestley 1775–1777, Contents). He referred to Joseph Black's discovery that the combination with fixed air *rendered* calcareous substances mild (1775–1777, vol. 1, 3). Coming to phlogiston, he spoke routinely of how air was transformed by the addition of phlogiston in a "regular gradation from dephlogisticated air, through common air, and phlogisticated air, down to nitrous air" (Priestley 1775, 392); this series was not an idle piece of theorizing, but something rooted in his daily laboratory operations. In his numerous experiments Priestley took note of all sorts of properties that substances took on as they were modified, but weight was not something he noted very frequently. Changes of weight seemed capricious in relation to phlogistication, which meant that weight was not a reliable or useful variable if one wanted to discern stable patterns in these phenomena.

What kind of experimental practices were linked to Lavoisier's thinking? He was also driven by pneumatic chemistry, but his fascination was about how air was absorbed and given out by solid substances in certain chemical reactions. That, combined with his quantitative bent, resulted in a focus on weight as a chief variable allowing the tracing of what came in and out of chemical substances, resulting in the "balance-sheet" method (Poirier 2005),[32] or the "algebraic" inclination (Kim 2005). Having taken up the compositionist way of thinking, Lavoisier sought and developed those experiments conforming to the classic compositionist practice of decomposition and recomposition, which was the most convincing method of ascertaining compositions. It was typically assumed that chemical components had weight, which could be tracked precisely using the balance, although it is not logically necessary for compositionism to take weight as the primary measure of the quantity of matter (for example, in mid-nineteenth-century organic chemistry, gas volume was often the most useful measure of the number of atoms involved in a reaction). A powerful compositionist system of theoretical and experimental practices grew up in Lavoisierian chemistry, and that is what supported the conviction that weight was the most important variable in chemical reactions. But those not operating in that system of practices would have found it difficult to share that conviction.

4. CONCLUSION, AND BROADER IMPLICATIONS

There was insufficient semantic incommensurability in the Chemical Revolution to prevent a reasonably straightforward theory choice. However, there was a significant degree of methodological incommensurability, at least on three dimensions: problem field, epistemic values, and practice-based metaphysics. This methodological incommensurability, not semantic incommensurability, is what made the Chemical Revolution such an underdetermined case. As Kuhn memorably put it in the middle of SSR (1970, 159): "lifelong resistance . . . is not a violation of scientific standards. . . . Though the historian can always find men—Priestley, for instance—who were unreasonable to resist for as long as they did, he will not find a point at which resistance becomes illogical or unscientific." Why exactly Kuhn thought Priestley was unreasonable in his resistance is not entirely clear to me. But I hope I have been successful in filling out Kuhn's verdict that Priestley (perhaps along with many others who defended phlogiston) did not at any point violate any superparadigmatic scientific standards, but merely had incommensurable scientific standards from Lavoisier and his colleagues.

I do not wish to claim that this one case can demonstrate any facts about incommensurability in general. However, I think the Chemical Revolution does make a powerfully suggestive case for us, as it did for Kuhn originally. If scientists in different paradigms are communicating with each other fairly well, which they usually do in scientific debates, then it is not likely that there is such a great threat posed by semantic incommensurability. On the other hand, the case of the Chemical Revolution illustrates how easily (and subtly) scientists in different paradigms can reach mutually irreconcilable judgments due to methodological differences. These differences could potentially create the threat of underdetermination, relativism, and irrationality.

My sense is that the worries about these threats got pinned on a misidentified target, namely semantic incommensurability, which is not so common in a serious form and easily defused when it does occur. Perhaps it is understandable that Kuhn himself maintained a focus on semantic incommensurability—by his own famous account, his thinking about incommensurability all began with the semantic aspect as he wrestled with the initial incomprehensibility of Aristotle's texts on physics and cosmology.[33] But I would argue that semantic incommensurability is generally more of a problem for the historians looking back than it is for the scientific actors at the time of theory change. The scientists in different paradigms can generally understand each other well enough, especially because most of the early proponents of the new paradigm will have been educated in the old. In the case of the Chemical Revolution the defenders of the old paradigm also understood the new paradigm well enough. The vexation of underdetermined theory choice came not from the shifts in meaning and reference, but from a lack of agreement on how to evaluate the merits of the competing paradigms due to the lack of shared standards of judgment—that is to say, due to methodological incommensurability.

I think methodological incommensurability is widespread and can be difficult to address precisely because it is less obviously noticeable than semantic incommensurability when it occurs. So my closing call is "back to *Structure*"—back to the engagement with methodological incommensurability, which Kuhn started so startlingly and productively in SSR. In contrast, his later concept of incommensurability, focusing on the semantic dimension (and a narrowly referential notion of semantics at that), is more precise yet less productive because it is more removed from scientific practice. My own view is that methodological incommensurability is not inherently harmful; rather, it forms the basis of pluralism in science, which I regard as a beneficial thing. But that is a subject for another day.[34]

NOTES

1. The most convenient place to examine Kuhn's later conception of incommensurability is Kuhn (2000, Chapters 1, 4, 11). For an authoritative secondary account see relevant sections of Hoyningen-Huene (1993) and, more briefly, Hoyningen-Huene (1990).
2. Hoyningen-Huene and Sankey note that Feyerabend agreed with this methodological thesis as well, but did not put it under the rubric of incommensurability, reserving the latter only for the semantic thesis. They also note a "perceptual" dimension to Kuhn's concept of incommensurability (ix), but do not enter into a detailed discussion of it.
3. This may not seem very "symbolic," but it is as symbolic as Kuhn's own example of "elements combine in constant proportions by weight."
4. For further details, see Chang (2012, Chapter 1).
5. Margolis (1993, 49; see also 44). Nonetheless, Margolis (1993, 50) says that the Chemical Revolution was a Kuhnian revolution, as the transition was "cognitively difficult though logically not so, hence best understood as turning on the presence of a barrier habit of mind."
6. See Musgrave (1976, 197) on acidity.
7. My expression is deliberately ambiguous here: some phlogistonists, following Henry Cavendish's 1766 version of the theory, regarded inflammable air as phlogiston itself. Later Cavendish (1784) modified his view to say that inflammable air was phlogisticated water, a formulation which Priestley also followed from then on. Musgrave (1976) gives a succinct summary of these conceptions.
8. Klein and Lefèvre (2007, 185; more generally, Chapter 10).
9. See Musgrave (1976, 190) for the description of Cavendish's 1766 version of phlogiston theory, and pages 201 and 203–205 for the 1784 version.
10. Such retrospective identifications would have horrified Kuhn, but the point about commensurability stands.
11. For more detail on these connections between phlogiston and concepts of post-Lavoisierian chemistry, see Chang (2009) and Chang (2011b).
12. On these chemists who admitted oxygen and also kept phlogiston, see Partington and McKie (1937–1939), Hufbauer (1982), and Allchin (1992).
13. See also Hoyningen-Huene (2008, 112).
14. On the early Kuhn's emphasis on this aspect of incommensurability, see Hoyningen-Huene (1993, 208–209).
15. Among philosophical commentaries on the Chemical Revolution, Andrew Pyle's (2000, 104) is one of the exceptional works that stress the importance of this factor.
16. See the very first chapter of Lavoisier ([1789] 1965). Some individual phlogistonists were also deeply concerned about heat and changes of state (e.g., Cavendish on boiling and evaporation), but these thoughts did not tend to affect their chemical theories. An apparent exception to the last statement is Jean-André De Luc, whose opposition to Lavoisier's chemistry seems to have been motivated by his ideas regarding the theory of rain; however, De Luc's chemistry was idiosyncratic and it is difficult to classify him as a phlogistonist.
17. This is part 2 (of 3) of Lavoisier's text ([1789] 1965, 173–291).
18. Priestley explained that this was "that which remains after exposing turbith mineral to a red heat"; the modern editor of his text adds that "turbith mineral" is "basic mercuric sulphate."
19. This passage occurs in a report to the Paris Academy by Lavoisier, Berthollet, and Fourcroy in 1787, which Lavoisier quotes in his comments on Kirwan's treatise on phlogiston; see Kirwan (1789, 15).

20. Chang (2009), section 2.
21. Cavendish added that there was one other consideration in addition to this prudence or conservatism: "it is more reasonable to look for great variety in the more compound than in the more simple substance," in relation to plants and their composition.
22. Lavoisier quoted in Donovan (1988, 219–220). Citing J. B. Gough and I. B. Cohen, Donovan goes on to caution that Lavoisier was being "less exact and less novel in anticipating a revolution in chemistry than has heretofore been assumed"; however, he by no means denies that Lavoisier had a "program of research" that was to "enable him to carry the study of chemistry to new heights."
23. Priestley ([1796] 1969, 17–18). He signed off his dedication with unflagging loyalty to the cause of the French Revolution: "I earnestly wish success to the arms of France, which has done me the honour to adopt me when I was persecuted and rejected in my native country. With great satisfaction, therefore, I subscribe myself Your fellow-citizen, Joseph Priestley."
24. See Bakewell (2010, 43, and Chapter 7) for a nice exposition of this aspect of Montaigne's thought. Perhaps there is a fruitful comparison to be made between Montaigne's *Essays* and Priestley's *Experiments and Observations on Different Kinds of Air*. A Cartesian quest for certainty is clearly eschewed in both.
25. This passage is from the preface to the French translation of Kirwan's treatise on phlogiston (Kirwan 1789, xiii).
26. Kirwan devoted the first chapter of his book on phlogiston (1789) to the consideration of weights. Cavendish was unsurpassed in his precision measurements of all things, including weights and the density of gases.
27. The term "principalist" has been commonly used, but I think it is more correct to spell it as "principlist," as we are referring to principles, not principals.
28. See Chang (2011a) for further details. I am drawing from the well-established work of many historians of chemistry, although the exact terminology of compositionism vs. principlism is my own. See, for example, Siegfried and Dobbs (1968), Siegfried (1982), Klein (1994), and Siegfried (2002). Klein (1996) clearly demonstrates the interaction between the experimental and the theoretical.
29. There are some very old echoes in principlism, including the old metaphysics of the substratum of elements modified by the influence of principles, even of matter being given form.
30. I am using Lavoisier's own figures given in his commentary on Kirwan (1789, 16), except that his pre-Revolutionary "g" was "grain," not the metric "gramme" that he himself helped to put in place.
31. Buchwald (1992) provides an instructive illustration of the instrumentation–ontology link, in a late-Kuhnian vein.
32. Thinking of the balance sheet of weights in chemical reactions must have been pleasing to the commercial–bourgeois sensibilities of Lavoisier's middle-class scientific community. But the oxygenists did not all share the same class background, and I have no means of supporting a real causal link here in any case.
33. See Kuhn (1999, 33) for one of his recollections of that formative moment.
34. See Chang (2012, Chapter 5).

REFERENCES

Allchin, Douglas. 1992. "Phlogiston after Oxygen." *Ambix* 39: 110–116.

Andersen, Hanne, Peter Barker, and Xiang Chen. 2006. *The Cognitive Structure of Scientific Revolutions*. Cambridge: Cambridge University Press.

Bakewell, Sarah. 2010. *How to Live: A Life of Montaigne in One Question and Twenty Attempts at an Answer*. London: Chatto & Windus.

Buchwald, Jed. 1992. "Kinds and the Wave Theory of Light." *Studies in History and Philosophy of Science* 23: 39–74.

Cavendish, Henry. 1784. "Experiments on Air." *Philosophical Transactions of the Royal Society* 74: 119–153.

Chang, Hasok. 2009. "We Have Never Been Whiggish (about Phlogiston)." *Centaurus* 51: 239–264.

Chang, Hasok. 2011a. "Compositionism as a Dominant Way of Knowing in Modern Chemistry." *History of Science* 49: 247–268.

Chang, Hasok. 2011b. "The Persistence of Epistemic Objects through Scientific Change", *Erkenntnis* 75: 413–429.

Chang, Hasok. 2012. *Is Water H_2O? Evidence, Realism and Pluralism*. Dordrecht: Springer.

Donovan, Arthur. 1988. "Lavoisier and the Origins of Modern Chemistry." *Osiris*, 2nd series, 4: 214–231.

Hoyningen-Huene, Paul. 1990. "Kuhn's Conception of Incommensurability." *Studies in History and Philosophy of Science* 21: 481–492.

Hoyningen-Huene, Paul. 1993. *Reconstructing Scientific Revolutions: Thomas S. Kuhn's Philosophy of Science*. Chicago: University of Chicago Press.

Hoyningen-Huene, Paul. 2008. "Thomas Kuhn and the Chemical Revolution." *Foundations of Chemistry* 10: 101–115. An earlier version was published in 1998 in V. M. Abrusci et al. (eds.), *Prospettive della logica e della filosofica della scienza*.

Hoyningen-Huene, Paul, and Howard Sankey (eds.) 2001. *Incommensurability and Related Matters*. Dordrecht: Kluwer, 2001.

Hufbauer, Karl. 1982. *The Formation of the German Chemical Community (1720–1795)*. Berkeley and Los Angeles: University of California Press.

Kim, Mi Gyung. 2005. "Lavoisier, the Father of Modern Chemistry?" In Marco Beretta (ed.), *Lavoisier in Perspective*. Munich: Deutsches Museum, 167–191.

Kirwan, Richard. 1789. *An Essay on Phlogiston and the Constitution of Acids*. New ed. London: J. Johnson. Originally published in 1784; French translation in 1788, with critical notes by Lavoisier and his colleagues; new English edition in 1789 by William Nicholson, with an English translation of the French notes, and Kirwan's replies. There is a modern reprint of the 1789 edition (London: Frank Cass & Co. Ltd., 1968).

Kitcher, Philip. 1978. "Theories, Theorists, and Theoretical Change." *Philosophical Review* 87: 519–547.

Kitcher, Philip. 1983. "Implications of Incommensurability." In Peter D. Asquith and Thomas Nickles (eds.), *PSA 1982*, vol. 2, 689–703.

Kitcher, Philip. 1993. *The Advancement of Science: Science without Legend, Objectivity without Illusions*. New York and Oxford: Oxford University Press, 1993.

Klein, Ursula. 1994. "Origin of the Concept of Chemical Compound." *Science in Context* 7(2): 163–204.

Klein, Ursula. 1996. "The Chemical Workshop Tradition and Experimental Practice: Discontinuities within Continuities." *Science in Context* 9(3): 251–287.

Klein, Ursula, and Wolfgang Lefèvre. 2007. *Materials in Eighteenth-Century Science*. Cambridge, MA: The MIT Press.

Kuhn, Thomas S. 1970. *The Structure of Scientific Revolutions*. 2nd ed. Chicago: University of Chicago Press.

Kuhn, Thomas S. 1977. "Objectivity, Value Judgment, and Theory Choice." In T. S. Kuhn, *The Essential Tension: Selected Studies in Scientific Tradition and Change.* Chicago: University of Chicago Press, 320–339.

Kuhn, Thomas S. 1983. "Commensurability, Comparability, Communicability." In Peter D. Asquith and Thomas Nickles (eds.), *PSA 1982*, vol. 2, 669–688.

Kuhn, Thomas S. 1999. "Remarks on Incommensurability and Translation." In Rema Rossini Favretti, Giorgio Sandri, and Roberto Scazzieri (eds.), *Incommensurability and Translation: Kuhnian Perspectives on Scientific Communication and Theory Change.* Cheltenham: Edward Elgar, 33–37.

Kuhn, Thomas S. 2000. *The Road Since Structure: Philosophical Essays, 1970–1993, with an Autobiographical Interview.* Ed. James Conant and John Haugeland. Chicago: University of Chicago Press.

Lakatos, Imre, and Alan Musgrave, (eds.) 1970. *Criticism and the Growth of Knowledge.* Cambridge: Cambridge University Press.

Lavoisier, Antoine-Laurent. [1789] 1965. *Traité élémentaire de chimie.* Paris: Cuchet, 1789. English translation, *Elements of Chemistry*, by Robert Kerr (1790), reprinted with an introduction by Douglas McKie (New York: Dover, 1965).

Lewis, Gilbert Newton. 1926. *The Anatomy of Science.* New Haven: Yale University Press.

Margolis, Howard. 1993. *Paradigms & Barriers: How Habits of Mind Govern Scientific Beliefs.* Chicago and London: University of Chicago Press.

Musgrave, Alan. 1976. "Why Did Oxygen Supplant Phlogiston? Research Programmes in the Chemical Revolution." In C. Howson (ed.), *Method and Appraisal in the Physical Sciences.* Cambridge: Cambridge University Press, 181–209.

Nordmann, Alfred. 1986. "Comparing Incommensurable Theories." *Studies in History and Philosophy of Science* 17: 231–246.

Odling, William. 1871. "On the Revived Theory of Phlogiston." (Address at the Royal Institution, April 28, 1871). *Proceedings of the Royal Institution of Great Britain* 6 (1870–1872): 315–325.

Partington, J. R., and Douglas McKie. 1937–1939. "Historical Studies on the Phlogiston Theory." *Annals of Science* 2 (1937): 361–404; 3 (1938): 1–58, 337–371; 4 (1939): 113–149 (in four parts).

Poirier, Jean-Pierre. 2005. "Lavoisier's Balance Sheet Method: Sources, Early Signs and Late Developments." In Marco Beretta (ed.), *Lavoisier in Perspective.* Munich: Deutsches Museum, 69–77.

Priestley, Joseph. 1775. "An Account of Further Discoveries in Air, in Letters to Sir John Pringle, Bart. P.R.S. and the Rev. Dr. Price, F.R.S." *Philosophical Transactions of the Royal Society of London* 65: 384–394.

Priestley, Joseph. 1775–1777. *Experiments and Observations on Different Kinds of Air.* 2nd ed., 3 vols. London: J. Johnson.

Priestley, Joseph. [1796] 1969. *Considerations on the Doctrine of Phlogiston, and the Decomposition of Water (and Two Lectures on Combustion, etc., by John MacLean).* New York: Kraus Reprint Co., 1969. This is a reprint of the original edition published in 1796 in Philadelphia by Thomas Dobson.

Pyle, Andrew. 2000. "The Rationality of the Chemical Revolution." In Robert Nola and Howard Sankey (eds.), *After Popper, Kuhn and Feyerabend.* Dordrecht: Kluwer, 99–124.

Sankey, Howard. 1991. "Incommensurability, Translation and Understanding." *The Philosophical Quarterly* 41: 415–426.

Siegfried, Robert. 1982. "Lavoisier's Table of Simple Substances: Its Origin and Interpretation." *Ambix* 29: 29–48.

Siegfried, Robert. 2002. *From Elements to Atoms: A History of Chemical Composition.* (*Transactions of the American Philosophical Society* 92: 4). Philadelphia: American Philosophical Society.
Siegfried, Robert, and Betty Jo Dobbs. 1968. "Composition: A Neglected Aspect of the Chemical Revolution." *Annals of Science* 24: 275–293.

Part III

Implications

9   Scientific Concepts and Conceptual Change

Hanne Andersen

Throughout his academic career, from the Lowell Lectures he gave at the age of twenty-nine and until the unfinished book manuscript that he left at his death, Kuhn struggled to develop a consistent account of scientific concepts and conceptual change. As his ideas developed he drew inspiration from many different sources, although often in a loose fashion that occasionally frustrated his critics.

This chapter will give an overview of the development of Kuhn's position, but it will discuss especially its affinity to the work of others and provide a discussion of the main issues that Kuhn left open for further development. The development of his position will be followed from his reflections on science education that informed his work leading up to *The Structure of Scientific Revolutions* (SSR), to his interest in family resemblance displayed in SSR and later works, to his focus on lexicons and lifelines in his last writings. The chapter will describe how Kuhn gradually developed a family resemblance account of concepts that in many respects is similar to accounts developed by cognitive scientists, and it will be shown both how the account can be further extended by drawing on additional work from the cognitive sciences and how such extensions can link the account more closely to philosophical discussions on such topics as experiments, explanation, robustness, and realism.

PRIOR TO SSR: SCIENCE EDUCATION

During the late 1950s, when Kuhn started wrestling with the ideas about normal science and paradigms that came to figure prominently in SSR, observations about the nature of science education played an important role in the development of his argument. Thus, in some of his very first presentations of his developing ideas he turned to an investigation of science education rather than to historical examples in order to substantiate the claims he advanced on the nature of science and its development. The first paper on the topic—"The Essential Tension: Tradition and Innovation in Scientific Research" (Kuhn 1959)—was delivered at the Third University of Utah Research Conference on the Identification of Scientific Talent in 1959. Here his views on the necessity of convergent thought for the activity of normal science were seen as appallingly conservative by the participating science educators, who were at the time interested in creativity rather than conformity.[1]

In "The Essential Tension" Kuhn described how during science education novices would first be presented with concrete problem solutions that the profession recognizes as paradigms and then asked to solve a number of additional problems closely related in both method and substance in order to learn to recognize categories of problem situations from the displayed exemplars. This was the first rudimentary form of his theory of concepts and classification, which at first received little development in SSR. The notion of concepts did not have any prominent position in the first edition of the monograph; instead Kuhn discussed more broadly the relation between rules, paradigms, and normal science. Thus, the basis for Kuhn's argument in SSR was his own experience as a historian that "the search for rules [is] both more difficult and less satisfying than the search for paradigms" (Kuhn 1970, 43). What could be disclosed by historical investigation of a given specialty at a given time was a set of standard *illustrations* of the application of various theories, but not a standard reduction of the standard applications to sets of *rules*.

Kuhn saw this emphasis on exemplars rather than rules as rooted in science education. During their education "scientists . . . never learn concepts, laws, and theories in the abstract and by themselves. Instead, these intellectual tools are from the start encountered in a historically and pedagogically prior unit that displays them with and through their applications" (Kuhn 1970, 46–47). And Kuhn found little reason to believe that scientists trained in this way would later develop abstract rules, once they had learned to identify scientific problems by resemblance to exemplars. To support this belief, Kuhn noted that "though many scientists talk easily and well about the particular individual hypotheses that underlie a concrete piece of current research, they are little better than laymen at characterizing the established bases of their field, its legitimate problems and methods" (Kuhn 1970, 47–48). Thus, it was a general claim about scientific practice *in toto* when Kuhn maintained that what the research problems within a given discipline "have in common is not that they satisfy some explicit or even some fully discoverable set of rules and assumptions" (Kuhn 1970, 45–46) but that, instead, they relate merely by *resemblance*.

Although reflections of science education played an important inspirational role for the early development of Kuhn's view, he never engaged any further with the topic. But later, other researchers in science education drew inspiration from Kuhn's work.[2] During the 1980s and early 1990s, several scholars argued that conceptual divides of the same kind as described by Kuhn's incommensurability thesis may in some cases exist in science education between teacher and student, and that science teaching should address these misconceptions in an attempt to induce conceptual change. Part of this research followed the thesis that cognitive ontogeny recapitulates scientific phylogeny,[3] and especially for the field of mechanics, research was done to show that children's naïve beliefs parallel early scientific beliefs. However, most research went beyond such identifications of analogies between students' naïve views and historically held beliefs. Instead, they investigated the historical records of the cognitive processes employed by scientists in constructing scientific concepts and theories more generally, focusing on the kinds of reasoning strategies employed in the construction and change of scientific concepts (see Nersessian 1992, 1995). Thus, this work still assumed that the cognitive activities of scientists in their construction of new scientific concepts was relevant to learning, but it saw the history of science merely as a repository of knowledge about how scientific concepts are constructed and changed, and it moved away from the Kuhnian emphasis on incommensurability and gestalt switch–like conceptual change.

AFTER SSR: FAMILY RESEMBLANCE

From SSR onwards, family resemblance became one of the central ideas from which Kuhn gradually developed his account of concepts during the following decades. Focusing on concept acquisition and the transmission of concepts from experts to novices, Kuhn explained how a novice learns a concept by being guided through a series of encounters with objects that highlight the relations of similarity and difference between the instances of the set of contrasting concepts. He illustrated his argument with an example of a very simple transmission of concepts: a child learning to distinguish waterfowl. In this example, the child is presented with various instances of ducks, geese, and swans, being told for each instance which category it belongs to. Further, the child is encouraged to try to point out instances of the concepts and will be corrected when making mistakes and gain praise when ascribing instances to the correct categories until, after a number of encounters, the child has acquired the ability to identify ducks, geese, and swans as competently as the instructor. The lesson that Kuhn drew from this example was that concepts are transmitted from one generation to the next by extracting similarities and differences between the exemplars on exhibit.

Based on this simple example, Kuhn assumed that the same technique applied for learning scientific concepts. But rather than learning to categorize individual objects, what one learns in the abstract sciences by this technique is to categorize problem situations and to apply laws to these different categories of problem situations. A law should be understood here as a law *sketch* or law *schema* whose detailed expression varies for different applications. For example, the law sketch F=ma, Newton's second law of motion, applies to problem situations involving a free fall in the form $mg=md^2s/dt^2$, to problem situations involving a simple pendulum in the form $mg \cdot \sin\theta = -ml \cdot d^2\theta/dt^2$, etc.

Thus, in learning scientific concepts the student is presented with a variety of problems which can be described by various expressions of a symbolic generalization. In this process, the student discovers a way to see each problem as *like* a previously encountered problem. Recognizing the resemblance, the student "can interrelate symbols and attach them to nature in the ways that have proved effective before. The law sketch, say f=ma, has functioned as a tool, informing the student what similarities to look for, signaling the gestalt in which the situation is to be seen" (Kuhn 1970, 189). Thus, a conceptual structure is established by grouping problem situations into similarity classes corresponding to the various expressions of the law sketch. As Kuhn stated it: "The resultant ability to see a variety of situations as like each other . . . is, I think, the main thing a student acquires by doing exemplary problems . . ." (Kuhn 1970, 189). Whether the concepts to be learned are simple concepts like the waterfowl "ducks," "geese," and "swans" or categories of complex problem situations involving law schemata like Newton's second law, the important point is that the grouping is not determined by necessary and sufficient conditions, but by family resemblance between the instances.

Kuhn later changed his view, separating concepts like "mass" and "force" from pure family resemblance concepts. Thus, in his later writing Kuhn introduced a distinction between *normic* concepts and *nomic* concepts. Normic concepts are "learned as members of one or another contrast set. . . . The ability to pick out referents for any of these terms depends critically upon the characteristics that differentiate its referents from those of the other terms in the set, which is why the terms involved must be learned together and why they collectively constitute a contrast set" (Kuhn 1993, 317). Nomic concepts, on the other hand, "stand alone. The terms with which it needs to be learned are closely related but not by contrast. . . . They are learned from situations in which they occur together exemplifying laws of nature" (317). It might seem that in this distinction between normic and normic concepts Kuhn saw the need to introduce a distinction between similarity class concepts and nonsimilarity class concepts. However, as argued by Andersen and Nersessian (2000), both normic and nomic concepts can be understood as similarity class concepts, but in the case of nomic concepts the family resemblances are among *complex problem situations* rather than among *individual objects* or phenomena.

Kuhn's family resemblance account of concepts is very similar to accounts developed by cognitive scientists as part of the Roschian revolution. During the 1970s, Rosch and other cognitive psychologists had conducted a range of experiments on people's use of everyday concepts like animals, trees, clothing, and furniture and had found that individual instances of a concept vary in how good an exemplar they are of the concept in question. From this empirical observation of these so-called graded structures they argued that a family resemblance account of concepts seemed to reflect the actual use of concepts much better than an account in terms of necessary and sufficient conditions.[4]

By the mid-1970s, Kuhn was in contact with Eleanor Rosch, and he was especially interested in her work on a special "basic level" in conceptual hierarchies. In his early work on family resemblance concepts Kuhn had briefly stated that his account "would have to allow for hierarchies of natural families with resemblance relations between families at the higher levels" (Kuhn 1970, 17n1), without making explicit what this would imply. However, one may note that Kuhn in his family resemblance account ascribed a special importance to *dissimilarity*, that is, the features which differentiate between instances of *contrasting* concepts, or contrast sets, as Kuhn called them. These are concepts whose instances are more similar to one another than to instances of other concepts and which may, therefore, be mistaken for each other. In this way the set of contrasting concepts form a family resemblance class at the superordinate level. On this analysis, family resemblance concepts form hierarchical structures in which a general concept decomposes into more specific concepts that may again decompose into yet more specific concepts, and so forth—in other words, taxonomies.

However, in his example of the child learning to recognize waterfowl, Kuhn had assumed that as a prerequisite to this learning process, the child had already learned to recognize birds (see e.g. Kuhn 1974, 309). Admittedly, this assumption solves the problem related to the "end point" of pointing, namely that in order for an ostensive act to be understood, the language learner must understand it not just as reference but as reference to something specific. One way of obtaining this is through prior determination of the domain out of which the pointing is to select something.[5] But this would lead to a regress since it presupposes that the superordinate concept is already known by the language learner, and in the end this would mean that language learning would have to start from the top of the hierarchy. However, starting with the work of Brown (1958), cognitive scientists had shown that language learning tends to start from intermediate levels of the conceptual hierarchies rather than from the top. Rosch and her collaborators followed up on this work in an effort to characterize this intermediate level of the conceptual hierarchy, which they termed the basic level. Thus, based on experiments that had shown that objects are more easily recognized as members of intermediate categories than as members of their super- or subordinate categories, they argued that categories at this intermediate level provide the best conceptual economy understood as the balance between maximizing information by using very specific concepts from the lower levels in the conceptual hierarchy, and minimizing cognitive effort by using very general concepts from the higher levels (Rosch, Mervis, Gray, Johnson, and Boyes-Braem 1976; Rosch 1978). Further, objects at this level also seemed to differ from higher levels in having very similar shapes, or by involving identical kinds of human interactions. Overall shape or ways of interaction could therefore be features that enabled language learners to start language learning at this intermediate level.

Kuhn admitted in a letter to Rosch that in developing his example on waterfowl, " . . . I just picked an arbitrary (?) level and resolved to leave problems of hierarchy aside until (much) later, rather assuming that nothing of much use conceptually would emerge in the process. That assumption was clearly wrong . . ." (Kuhn to Rosch, December 7, 1976). Kuhn never unfolded his thoughts on the basic level in his published work. However, it is clear from his unpublished material—for example, lecture notes from one of the last courses he taught at MIT before his retirement (Kuhn, Philosophy 24.853)—that he thought of basic level categories as categories that achieve a good compromise between maximizing similarity of their members and minimizing their similarity with members of other categories.[6] Similarly, Giere (1994) has argued that in, for example, classical mechanics a basic level can be identified as the level at which members of the category appear visually more similar to each other than to members of other categories at that level. This view has been vindicated by categorization studies, where it has been found that novices operate at this visually privileged basic level when sorting basic science problems, while experts have acquired the ability to categorize by higher and more abstract levels in the hierarchy (see e.g. Chi, Feltovich, and Glaser 1981).

EXTENDING KUHN'S WORK ON TAXONOMIES: EXPERIMENTS AND EXPLANATIONS

As Kuhn developed his account of scientific concepts, he increasingly emphasized the importance of contrast sets and the features useful for differentiating between them. However, he primarily analyzed everyday concepts and said very little about scientific concepts. Even in his unpublished works from the 1990s he still returned to his old favorite example of waterfowls, showing how a superordinate concept decomposes into a group of contrasting concepts, and how this decomposition is determined by sets of features (Figure 9.1).

However, the basic idea that "to each node in a taxonomic tree is attached a name . . . and a set of features useful for *distinguishing* among creatures at the next level down" (Kuhn 1990, 5) is obviously useful for many scientific concepts as well. Thus, Kuhn's idea of differentiae has been adopted and elaborated upon by, among others, Buchwald (1992) and Chen (1997), who have shown how instruments and experiments play an important role in distinguishing kinds. For example, Buchwald argues that "experimental work divides the elements of the [taxonomic] tree from one another: sitting at the nodes or branch-points of the tree, experimental devices assign something to this or that category (Buchwald 1992, 44). Similarly, Chen argues that "instruments also play an important role in establishing lexical taxonomies. . . . Instruments practically designate concepts in a lexical taxonomy by sorting their referents under different categories" (Chen

1997, 269). On this view, instruments are sorting devices which distinguish instances of contrasting concepts by determining specific features which differ for the contrasting concepts.

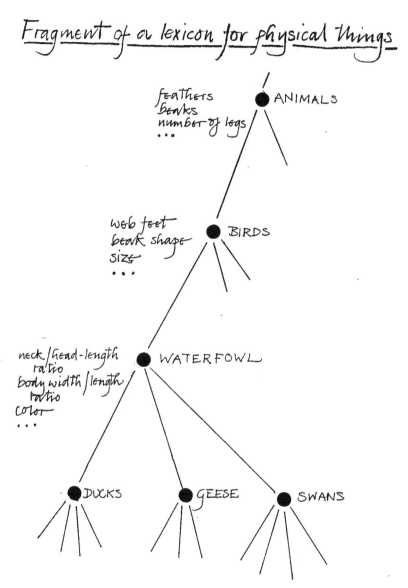

Figure 9.1 Fragment of a lexicon. The concept of waterfowl is decomposed into the concepts of ducks, geese, and swans by a number of differentiating features. *Source:* Thomas S. Kuhn, "An Historian's Theory of Meaning." Reprinted with permission of the MIT Institute Archives and Special Collections.

Another point that Kuhn did not pursue in much detail was the *correlation* of the set of features attached to each category. Admittedly, Kuhn emphasized that the use of a scientific concept may be governed by a number of different criteria, and that their coexistence represents knowledge of how the world behaves, about "the situation that nature does and does not present" (Kuhn 1970, 191). Thus, concepts are projectible in the sense that they imply hypotheses of how their instances behave, and this projectibility may develop gradually as new features are discovered.

Initially, a concept may be introduced just on the basis of a single differentiating feature that distinguishes its instances from instances of a known concept. As Buchwald has noted, "a novel taxonomy may emerge as someone attempts to grapple with a particular device" (Buchwald 1992, 44). For example, one of Kuhn's examples in SSR, the discovery of X-rays, describes how Roentgen one day noticed that a barium platino-cyanide screen at some distance from his cathode ray tube glowed when the discharge was in process and that this effect was due to some new agent.[7]

However, it usually takes more than just different behaviors of a device to posit new categories. Thus, Buchwald also added the qualifying notion of the "strength" or "robustness" of a taxonomy which to some extent reflects its device independence, "the ease with which it can be separated from the device" (1992, 44). Although concepts may be introduced on the basis of just a single differentiating feature, it is at the same time crucial that "a robust taxonomy is also compatible with many other devices that do what the taxonomy considers to be the same thing that the first one does but in entirely different ways" (44). Thus, a concept becomes increasingly entrenched as more and more features or combinations of features select the same category.

Although Kuhn never discussed this explicitly, he seemed intuitively to link something similar to this idea of entrenchment with the question of when to posit a new entity. Thus, in his account of the discovery of X-rays he asked the question at what point in Roentgen's investigation one could say that X-rays had been discovered. Kuhn discarded the view that the initial observation of the glowing screen would suffice. But he also discarded the view that the discovery occurred at the end of Roentgen's hectic weeks of research when he was exploring the properties of a new radiation that he had already discovered. For Kuhn, X-rays emerged at some point during the weeks during which Roentgen was investigating how the glow of a screen came in straight lines from the cathode ray tube, that the radiation cast shadows, that it could not be deflected by a magnet, etc.

More explicit considerations on when to posit a new entity have been advanced by Arabatzis (2008), who in his analysis of when scientists are ready to posit the existence of hidden entities argues that, for scientists, "the over-determination of a hidden entity's properties in different experimental systems is often an important reason in favor of its existence" (Arabatzis 2008, 14). Similarly, Gooding (1986) has noted with respect to new experimental possibilities, or, as he terms them, construals, that "until the significance of novel information has been sketched out, construals of it retain the provisional and flexible character of possibility. . . . The fact that an array of effects can be construed tends to support their facticity, and lends credibility to the construal" (Gooding 1986, 219).

But there is more to robustness than just device independence and correlation of features. Although initial, explorative research in a new area may often be focused on empirical examination of possible correlations of features—which features seem to be correlated, which new concepts based on these feature correlations arise, and so on—what usually follows is the development of *reasons* for these correlations, *theories* that explain why specific features are correlated.

The urge to derive theories explaining the correlation of features is not specific for scientific concepts, but has been the topic of general discussions in cognitive science. Cognitive scientists such as Murphy and Medin (1999) have argued that, generally, people tend to deduce causal explanations for feature correlations, a view that has become known as the Theory-Theory of concepts. This has occasionally been interpreted as a return to essentialism (e.g., Laurence and Margolis 1999, 47), which may seem counter to Kuhn's antiessentialist, family resemblance account of concepts. However, it is important to note that although the focus on underlying causal explanations of the correlation of surface attributes might seem to encourage essentialist views, theory-theorists have emphasized that the features which appear essential do not do so because of the structure of the world, but because they are the features that are most central to our current understanding of the world (Murphy and Medin 1999, 454). Later versions of the Theory-Theory point out that people tend to give more weight to features that are seen as the causes of other features (Ahn 1998, 138; Sloman, Love, and Ahn 1998), at the same time emphasizing that this view is different from essentialism in that causal features need not be defining features, and by the same token that features need not be dichotomized into essential and nonessential features. Thus, the Theory-Theory may be seen as an important addition to Kuhn's account rather than as a challenge to it.

SKETCHING THE DYNAMICS: THE NO-OVERLAP PRINCIPLE

Kuhn's account of hierarchical conceptual structures was developed gradually over several decades in an attempt to explain incommensurability as a phenomenon observed by historians of science. When Kuhn started elaborating on the issue of conceptual hierarchies in the late 1980s and early 1990s, he also introduced a number of restrictions that concepts in a lexicon have to obey. The restriction that figured most prominently in his writings was the "no-overlap principle": "No two kind terms, no two terms with the kind label, may overlap in their referents unless they are related as species to genus. There are no dogs that are also cats, no gold rings that are also silver rings, and so on; that's what makes dogs, cats, silver, and gold each a kind" (Kuhn 1991, 4). Kuhn did not spell out what the other restrictions were, but just indicated that further restrictions existed. However, classification and the division of concepts has always been an important part of classical logic, and it is common to characterize taxonomic divisions by the three principles that concepts formed by a division (1) do not overlap, (2) are subordinates to the same superordinate concept, and (3) together exhaust the superordinate concept.[8]

A predecessor to the no-overlap principle can be found much earlier in his work on family resemblance concepts, as a condition which the world must meet for his account of concepts to be possible at all. Thus, early on he had argued that if categories could overlap, then it would be necessary to define their extensions. As a simple *modus tollens* argument he added that since it can be observed that the extensions of categories are not defined, then they cannot overlap: "Only if the families we named overlapped and merged gradually into one another—only, that is, if there were no *natural* families—would our success in identifying and naming provide evidence for a set of common characteristics corresponding to each of the class names we employ" (Kuhn 1970, 45). Thus, it had been a key feature of Kuhn's family resemblance account of concepts from the outset that it depends on an "empty perceptual space between the families to be discriminated" (Kuhn 1970, 197n14). In formulating this condition on the world, Kuhn explicitly moved beyond Wittgenstein, whom he found to have said "almost nothing about the sort of world necessary to support the naming procedure he outlines" (Kuhn 1970, 45n2).

This condition on the world that it has to offer empty perceptual spaces between the families to be discriminated is also the key to understanding Kuhn's position in-between traditional realism and traditional antirealism. When Kuhn claimed that "though the world does not change with a change of paradigm, the scientists afterwards work in a different world" (Kuhn 1970, 121), he distinguished between two different kinds of worlds: a "hypothetical fixed nature" (Kuhn 1970, 118) and a "perceived world" (Kuhn 1970, 128). Like Kant's thing-in-itself, Kuhn saw the hypothetical fixed nature as "ineffable, undescribable, undiscussible," as a "Kantian source of stability," "located outside of space and time" (Kuhn 1991, 12). The perceived world, on the other hand, is "a world already perceptually and conceptually subdivided in a certain way" (Kuhn 1970, 129), and this subdivision was a structure that was imposed on the world by means of the concepts applied to it. Because this perceptually and conceptually subdivided world was dependent on conceptual structures and would therefore change with a change of language, Kuhn maintained that his view was Kantian but "with categories of the mind which could change with time as the accommodation of language and experience proceed" (Kuhn 1979, 418–419). A decade later, he elaborated on this comparison, now directly comparing Kant's categories with the lexicon: "Like the Kantian categories, the lexicon supplies preconditions of possible experience. But lexical categories, unlike their Kantian forebears, can and do change, both with time and with the passage from one community to another" (Kuhn 1991, 12).

This position may seem to contain an inner tension: on the one hand, the structure of perceptual space is determined by relations of similarity and difference between instances of contrasting concepts, but on the other hand these relations are dependent on a certain structure of the perceptual space. To dissolve this tension it is important to note that the position that Kuhn was developing was inherently historical. On Kuhn's view, the phenomenal world is never structured from scratch by its inhabitants. Instead, they will always have inherited a phenomenal world from their predecessors which they can then interact with and change in the process. In this way, every new generation will find a phenomenal world

> already in place, its rudiments at their birth and its increasingly full actuality during their educational socialization, a socialization in which examples of the way the world is play an essential part. . . . Creatures born into it must take it as they find it. They can, of course, interact with it, altering both it and themselves in the process, and the populated world thus altered is the one that will be found in place by the generation which follows. (Kuhn 1991, 10)

Hence, a structured perceptual world populated with discriminable families is always inherited by any generation from their predecessors. But once the new generation has gained access to this particular phenomenal world they may start reshaping it by introducing new relations of similarity and difference and abandoning old ones and thus leave to their successors a structure of the phenomenal world different from the structure they inherited themselves.

Kuhn's position is realist in the sense that nature cannot be forced into any arbitrary set of conceptual boxes. Whether a set of features really is correlated in the sense that they all distinguish coextensive categories was for Kuhn an objective matter. If they did not, anomalies would inevitably appear. But the claim that it is an objective matter whether features can be correlated in a particular way does not rule out that features can be correlated in different ways as well and that these will carve different joints in the phenomenal world. Thus, Kuhn's position should be understood as a purely negative claim that not any arbitrary correlation of features is possible; it is not a positive claim about the existence of a privileged set of features that carve out the world's real joints.

Whereas his early work had focused on empty perceptual space as a condition which the world must meet for a family resemblance account of concepts to be possible, his later work also focused on the function of this condition for the process of conceptual change. On Kuhn's view, these hierarchical restrictions are fundamental for a lexicon, and the lexicon will therefore be seriously challenged if an instance of one of the categories in the lexicon turns up that violates them. Again, Kuhn was primarily concerned with the no-overlap principle, and his own examples concerned discoveries that could be classified into contrasting categories: "if the members of a language community encounter a dog that's also a cat (or, more realistically, a creature like the duck-billed platypus) they cannot just enrich the set of category terms but must instead redesign a part of the taxonomy" (Kuhn 1991, 4). In this way, violations of the no-overlap principle (as well as the two other taxonomic principles) can be seen as the key to understanding what may provoke *changes* in a lexicon. However, although he elaborated on the notion of anomalies—especially in the form of violations of the no-overlap principle—as events that may *trigger* conceptual change, he did not say much about the creative process through which a new conceptual structure is created. This may in part be a reminiscence of the gestalt-switch metaphor which he had introduced in SSR and which seemed to indicate that conceptual changes happen all at once. Further, he had started to develop his account in an attempt to explain incommensurability as the relation between two conceptual structures on each side of a revolution and thus as the end *result* of conceptual change, while the *process* of scientific change did not come into his focus until much later.

During the 1980s Kuhn started admitting that his interpretation of revolutions and later of translation failure had been the result of his experiences as a historian discovering the past, and he realized that the experience of the scientists moving through time in the opposite direction would most probably be quite different. Admittedly, he still argued that scientists experience gestalt switches, but he now added the qualification that "their shifts in gestalt will ordinarily be smaller than the historian's for what the latter experiences as a single revolutionary change will usually have been spread over a number of such changes during the development of the sciences" (Kuhn 1983b, 715). However, he still denied that these language changes might have taken place originally as what he called "gradual linguistic drift." In support of his view he referred to empirical evidence in the form of reports of "aha" experiences (Kuhn 1983b, 715). Further, Kuhn argued that "it is the acceptance of fuzziness that permits drift, the gradual warping of the meanings of a set of interrelated terms over time" and that "in the sciences borderline cases of this sort are sources of crisis," and he concluded that although gradual linguistic drift may happen in discourses such as those of the political life, it is simply inhibited in the sciences (Kuhn 1983b, 715).

Kuhn never developed any detailed case studies on violations of the no-overlap principle, but several other scholars have conducted detailed historical case studies that vindicate the view that violations of the no-overlap principle may trigger conceptual change. Among these, Chen (2002), in a direct parallel to Kuhn's example of the duck-billed platypus, has analyzed the developments in nineteenth-century ornithology, when the discovery of new species on distant continents questioned the existing taxonomy for classification of birds. Focusing specifically on the discovery of the South American screamer whose webbed feet and pointed beak provided an overlap in a taxonomy that was based on a clear difference between water birds with webbed feet and rounded beaks and land birds with clawed feet and pointed beaks, Chen showed how this discovery led ornithologists to present alternative taxonomies designed to eliminate the anomalies caused by the newly discovered species. Likewise, Andersen (1996) has analyzed the development of nuclear physics in the 1930s leading up to the discovery of nuclear fission, showing how a series of anomalies led to gradual conceptual changes, and how in this process discoveries of anomalies in the form of violations of the no-overlap principle interact with different forms of conceptual development, including both the introduction of new differentiating features, the introduction of new concepts, and changes in the theoretical explanation of correlated features.

Several scholars have described typologies of the different kinds of conceptual changes that may result from anomalies (see e.g. Andersen, Barker, and Chen 2006; Thagard 1992). For example, anomalies may suggest that an instance of a given concept within the taxonomy behaves differently than expected and is therefore characterized by other differentiating features, but without changing the boundaries of the concepts in the taxonomy. Further, anomalies may suggest that a new concept should be added within the taxonomy, but simply as an additional category of previously undiscovered objects such that the new concept does not affect the boundaries of previously known concepts. Finally, and most severely, anomalies may suggest that the previously assumed conceptual boundaries do not hold, that is, that the taxonomy must be restructured in order to categorize all encountered objects consistently. Whereas the two former kinds are changes that can be assimilated within the existing taxonomic structure, the later kind changes the taxonomic structure itself. Although these studies provide typologies of conceptual change and accounts of how anomalies trigger changes and how a series of anomalies may drive a stepwise development of conceptual change, they do not amount to an account of the process through which the changes are brought about. However, recent research in cognitive history of science has uncovered many of the reasoning practices through which scientists construct new concepts, (see e.g. Nersessian 2008a, 2008b).

LAST SPECULATIONS: LIFELINES AND CATEGORIZATION MODULES

Kuhn's condition on the world that there has to be empty perceptual space between the families to be discriminated had been part of his position from early on. In some of his last writings he tried to sketch a justification of the

basic principles on which his account was based. Kuhn had often emphasized that his account of concepts applies only to a particular kind of concept, namely those "which refer to the objects and situations into which a
language takes the world to be divided" (Kuhn 1990, 4). On this view, a
taxonomic conceptual structure, or a lexicon as Kuhn also called it, is a
sort of categorizing module in which certain sorts of expectations about the
world are embedded (cf. Kuhn 1990, 5–8). In some of his last talks, Kuhn
argued that the lexicon should not be seen as a set of beliefs, but as "a mental module prerequisite to having beliefs, a mode that at once supplies and
bounds the set of beliefs it is possible to conceive" (Kuhn 1991, 5). Thus,
Kuhn hypothesized that "the underlying form of taxonomic classification
embodied in the lexicon is the technique required to the reidentification of
individuals" (Kuhn 1990, 13). The basic argument was that one of the fundamental requirements for human survival is the ability to track objects,
and this requires, first, that the lifelines of objects do not intersect (see
Figure 9.2), and second, that the lifelines can be traced by humans through
the features of the objects, in other words, that we "need features such that
for some minimal time interval, two successive presentations of an object

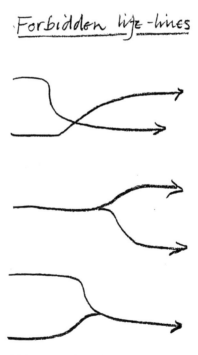

Figure 9.2   Forbidden lifelines. Lifelines do not intersect.
Source: Thomas S. Kuhn, "An Historian's Theory of Meaning." Reprinted with permission of
the MIT Institute Archives and Special Collections.

along the same life line will be more like each other than any presentation of an object on another life line" (Kuhn 1990, 13; see Figure 9.3).

Kuhn expected this mental module to be both biologically and culturally determined: "Doubtless some aspects of [the] lexical structure are biologically determined, the products of a shared phylogeny. But, at least among advanced creatures (and not just those linguistically endowed), significant aspects are determined also by education" (Kuhn 1991, 10).[9] The biological component made Kuhn reflect on the evolutionary origin of the module, claiming that "presumably it evolved originally for the sensory, most obviously for the visual, system" and that it developed from "a still more fundamental mechanism which enables individual living organisms to reidentify other substances by tracing their spatio-temporal trajectories" (Kuhn 1991, 5). From his teaching notes (Kuhn, Philosophy 24.853) he seemed to have drawn on standard psychological literature on infants' development of reidentification skills, such as Bower, *Development in Infancy* (1974), Keil, *Semantic and Conceptual Development* (1979), and Markman, *Categorization and Naming in Children: Problems of Induction* (1989). Based on

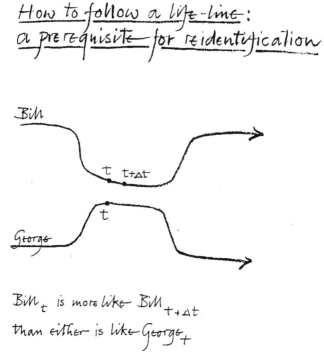

Figure 9.3 How to follow a lifeline. Two successive presentations of an object along the same lifeline will be more like each other than any presentation of an object on another lifeline.
Source: Thomas S. Kuhn, "An Historian's Theory of Meaning." Reprinted with permission of the MIT Institute Archives and Special Collections.

this psychological literature he described how reidentification by tracking with the eyes and hands could be seen in infants from the age of a few months and argued that when these children reacted to an object placed inside another as if it had disappeared, this could be seen as the root of the principle that lifelines do not intersect. Admittedly, this example concerned spatiotemporal overlap, but Kuhn argued that later infants would develop the capacity for tracking objects by qualities rather than just through spatiotemporal continuity, and that the principle that lifelines do not intersect would develop into a no-overlap principle for qualities. However, these reflections remained sketchy and did not draw in much detail on available resources from cognitive science. Most importantly, as Nersessian (2003) has pointed out, the psychological literature on which Kuhn was drawing concerned numerical identity and individuation of objects, and it was a category mistake to apply results on how individuals are tracked through time to the question of how concepts are tracked through time.

ROOMS FOR EXPLORATION: COMMUNITY AND INDIVIDUAL

When Kuhn realized the differences between conceptual shifts as experienced by the historian working backwards in time and the conceptual changes created by the scientists taking part in the scientific development, he adhered to the view that historians, working backwards, may experience only a single conceptual shift whereas the actual developmental process created by scientists working forwards in time may require a whole series of stages. However, it remained unclear from Kuhn's writings whether he thought that these minor changes could in addition be distributed among different members of the scientific community. In Kuhn's work conceptual structures are sometimes described as properties constitutive of a scientific community and sometimes as properties carried by each individual member of the community, but how these two views interrelate was never spelled out in any detail.

Ever since SSR, Kuhn had been interested in the relation between a scientific community and the conceptual structures that the members of this community share, but he had difficulties spelling out this relation. Initially, he had introduced scientific communities and the various cognitive elements that he at that point still referred to with the overarching concept of paradigms in a circular way, namely that "a paradigm is what the members of a scientific community share, and conversely a scientific community consists of men who share a paradigm" (Kuhn 1970, 176)—a circularity that he pointed out in the Postscript to the second edition of SSR.[10] Later, in his writings from the 1980s and 1990s, after he had taken his linguistic turn, Kuhn had replaced the diffuse notion of paradigms with the more specific notions of taxonomies and lexicons as the description of the cognitive resources that the members of a scientific community share.

He now described the members of a scientific community as members of a language community who through their language had similar expectations with respect to the objects and situations to be encountered. On this view, what members of a scientific community must share is the overall lexical structure that enables them to pick out the same referents for the same terms, but at the same time the features that they each use in categorization may vary:

> Homologous structures, structures mirroring the same world, may be fashioned using different sets of criterial linkages. What such homologous structures preserve, bare of criterial labels, is the taxonomic categories of the world and the similarity/difference relationships between them. . . . What members of a language community share is homology of lexical structure. Their criteria need not be the same, for those they can learn from each other as needed. But their taxonomic structures must match, for where structure is different, the world is different, language is private, and communication ceases until one party acquires the language of the other. (Kuhn 1983, 683)

This description was developed to explain incommensurability as a relation of untranslatability.[11] However, the process by which a new lexical structure was developed that was incommensurable with the old remained opaque and was characterized only through the negative claim that it could not be through gradual linguistic drift since that would require a fuzziness of the conceptual borders that science did not permit (cf. Kuhn 1983, 715).

Only in some of his last works did he explicitly address the microprocesses that occur during the process of conceptual change, noting that "as the conceptual vocabulary of a community changes, its members may undergo gestalt switches, but only some of them do and not all at the same time. . . . To speak . . . of a community's undergoing a gestalt switch is to compress an extended process of change into an instant, leaving no room for the microprocesses by which the change is achieved" (Kuhn 1989, 50). However, Kuhn added that his work had nothing to say about these microprocesses that occur within a community during periods of conceptual change, only that he saw his work as "designed to leave room for their exploration" (Kuhn 1989, 51).

One kind of microprocess that is related to the graded structures of family resemblance concepts has been examined by Andersen, Barker, and Chen (2006). As described above, Kuhn's account of concepts shares the implication with other family resemblance accounts that different instances of the same concept are not necessarily equally good examples of this concept. Since category membership is determined from the degrees of similarity to other instances of the concept and degrees of difference to instances of contrasting concepts, an instance which is similar to other instances of the concept with regard to many features may be seen as a better instance than one that shares only a few of these. Similarly, an instance that is similar to other instances with regard to features that figure prominently in theories explaining their correlation with other features may be seen as a better instance than one that displays features less central for theoretical explanations.

On Andersen, Barker, and Chen's reconstruction of Kuhn's position, the existence of graded structures can explain some of his key claims about the role played by anomalies in the development of science, especially that anomalies are not equally severe, and that different scientists may judge anomalies differently. Based on Kuhn's identification of anomalies with violations of the no-overlap principle, Andersen, Barker, and Chen argue that if an object is encountered that, judged from different features, seems to be an instance of two contrasting concepts, it violates the expectations regarding which objects exist and how they behave. If the encountered object judged from different features is a *good* example of two different concepts in a contrast set, this will be a severe anomaly, as it clearly violates the no-overlap principle and thereby calls the adequacy of the conceptual structure into question. On the contrary, if an object is encountered that, judged from different features, is a *poor* example of two different concepts in a contrast set, it may not call the conceptual structure into question, but just suggest that further research may be necessary to find out whether, for example, a new concept should be introduced or whether the existing concepts may be enriched with some additional features that allow the objects to be unequivocally assigned to one of them. Further, since the graded structures are based on feature-based similarities and differences and since different speakers may judge category membership using different features, different members of the scientific community may have different graded structures of their concepts and therefore have different views on which anomalies are severe and which are not. In such cases, only some will react to an anomaly by changing (parts of) their conceptual structure, while others will instead ignore the anomaly or draw on additional features to develop a more unequivocal classification into the known categories.[12]

This analysis only applies to communities that share taxonomic structures, that is, to communities whose members have gone through similar training in the same area of expertise. However, much science, not least much contemporary science, is developed by groups of scientists with different backgrounds and whose different areas of expertise complement each other. This is a situation that Kuhn's account does not in any way address. Kuhn's account assumes ethnocentrism of disciplines, that is, that scientific disciplines are clusters of specialties with clear gaps between them, and unidisciplinary competence, that is, that scholars are competent in one distinct discipline.[13] In contrast to this view it may be argued that scientific specialties often do overlap and that scientists in their collaborations often integrate knowledge drawn from multiple areas of expertise. Kuhn's account of scientific concepts provides no guidance to this situation, and the remaining part of this chapter will provide a brief sketch of what needs to be developed in order to extend Kuhn's account to cover interdisciplinary work.

Turning our attention away from scientists sharing conceptual structures to scientists with overlapping conceptual structures, new aspects of the social process of conceptual change introduced by groups of scientists are brought to the foreground. First, Kuhn's account needs to be extended so that taxonomies are interconnected in complicated crisscrossing patterns, because concepts are related to other concepts embedded in different lexicons, for example through regularities or laws.[14] Second, Kuhn's account also needs to be extended so that concepts and the knowledge that they entail can be distributed among the members of the scientific community in the sense that different members draw on different parts of the interrelated taxonomies.

A brief example may serve as illustration: research on induced radioactivity in heavy nuclei during the 1930s. An important part of this research was to bombard heavy nuclei with neutrons to see whether that resulted in radioactive decay, and, for beta-decay that increases the atomic number, whether elements beyond uranium in the periodic table could be produced. The two main groups working in this area were Fermi's team in Rome and the Meitner-Hahn-Strassmann team in Berlin. Both teams included both nuclear physicists and analytical chemists who worked together closely to examine the decay series and to identify the elements that were produced in the process. While nuclear physics provided knowledge about the likely decay processes, analytical chemistry provided knowledge about identification of the elements. And whereas physicists and chemists had some knowledge of each other's fields—physicists knowing the basics of the periodic system, chemists knowing the basics with regard to which disintegration processes to expect—there were also nonshared areas such as the nuclear physicists' detailed computations of tunneling effects or analytical chemists' detailed knowledge of precipitation processes.

In this research, several taxonomic structures are interconnected, most importantly the taxonomy of decay processes that includes α-emission, p emission, n capture, and β emission (shown in Figure 9.4 in white) and the taxonomy of the produced daughter elements that range in the periodic system from $Z-2$ to $Z+1$ compared to the mother nucleus Z (shown in Figure 9.4 in grey). Each decay process produces a daughter element that can be identified by its chemical characteristics, and this provides a close link between the two taxonomies (shown in Figure 9.4 as arrows).

During the period from 1934 to 1938 the two groups conducted long series of experiments, and one can follow step-by-step how new features and new concepts were introduced into the taxonomies, and how new theories were developed to explain discovered correlations.[15] But a major conceptual restructuring was made when the two chemists Hahn and Strassmann in December 1938 discovered that a decay product that they thought was

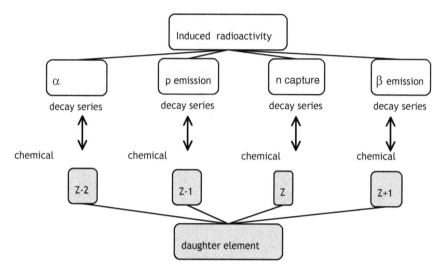

Figure 9.4 Interrelation between the taxonomy of decay products produced in induced radioactivity and the taxonomy of chemical elements. Figure by author.

radium could not be separated from barium in their precipitation process— for chemists a very clear violation of the no-overlap principle. Hahn wrote to their former group member, the physicist Lise Meitner, that although he knew this was impossible from a physical viewpoint, as a chemist he had to conclude that the element was barium and that the nucleus had split. Meitner discussed Hahn's report with another physicist, her nephew Otto Frisch, and together they realized that the resources were available to give a physical explanation of this new fission process. After this was communicated back to Hahn, Hahn and Strassmann published their result that the nucleus had split, while Meitner and Frisch published the physical explanation of this new phenomenon.

What is important in this brief case study is that the new concept of nuclear fission was not introduced by any individual person or within a single taxonomy. It was only introduced through the joint acceptance by the chemists Hahn and Strassmann and by the physicists Meitner and Frisch of an ontological claim about the existence of a new process, as it could be incorporated into the two interconnected taxonomies of decay products and chemical elements, respectively. Further, the introduction of the new concept of nuclear fission which had originally been triggered by what would otherwise have been a severe violation of the no-overlap principle in the taxonomy of chemical elements implied correlations of features in the taxonomy of decay products, and these correlations required new theoretical explanations. In this process of introducing a new concept, the physicists and chemists brought together various parts of the conceptual structures in which the new concept was embedded, parts that they each individually possessed, and which were jointly necessary to introduce the new concepts without violating any of the expectations entailed by the interconnected taxonomies describing this aspect of the world. Combining their individual knowledge, the involved scientists therefore had to mutually trust each other's expertise.

Several issues therefore have to be analyzed in order to understand conceptual developments in the intersection between overlapping conceptual structures. An extension of Kuhn's model to include the microprocesses occurring within a community of collaborating scientists during periods of conceptual change needs to include insights into how to combine distributed conceptual structures, into how to arrive at a joint consent to accept conceptual changes in the overlap between different conceptual structures, and into the relations of trust required to bridge partially overlapping conceptual structures.[16]

ACKNOWLEDGMENTS

I would like to thank K. Brad Wray, Theodore Arabatzis, and Vasso Kindi for valuable comments on earlier versions of this chapter. I would also like to thank the Danish Council for Independent Research * Humanities for supporting the project "Philosophy of Contemporary Science in Practice."

NOTES

1. For discussions of how Kuhn drew on science education in the development of his views as well as of early reactions, see Andersen (2000a).
2. For overviews of research on conceptual change in science education, including the inspiration drawn from Kuhn, see, for example, Duschl (1991), Linn (2008), Treagust and Duit (2008), Vosniadou (2007).
3. The thesis that ontogeny recapitulates phylogeny was originally advanced by Piaget, who saw intriguing parallels between conceptualizations of nature as they developed through the history of science and those produced during the cognitive development of the individual. Later scholars have differed from Piaget in using historical cases primarily to characterize various kinds of conceptual changes that also take place in learning and development rather than trying to fit the unfolding of history into a theory of cognitive development. See also Nersessian (1998).
4. Although there have not been empirical studies of possible graded structures of scientific concepts, Giere (1994) has suggested how the Roschian account of concepts can be applied to mechanical concepts, arguing that models of classical mechanics form categories with graded structures.
5. See Hoyningen-Huene (1993, 80) for a full account of this issue in relation to Kuhn's early work.
6. For a detailed account of Kuhn's family resemblance account of concepts and its relation to the work of Rosch and others, see Andersen (2000b). For accounts of the relation between Kuhn and Wittgenstein more generally, see Barker (1986), Cedarbaum (1983), and Kindi (1995), and for relations between Kuhn and the Roschian revolution, see Nersessian (1998).

7. Cf. Kuhn (1970, 57–58). See Andersen (2002, 2008) for further details on the conceptual developments during the early research on X-rays and radioactivity.

8. For a detailed account of these principles in relation to Kuhn's account of concepts, see Andersen, Barker, and Chen (2006, Chapter 4). Among the philosophical roots of analyses of conceptual hierarchies is Aristotle's treatment of definitions in terms of genus and differences, which in the scholastic tradition was developed into definitions *per genus proximum and differentiam specificam*. However, despite his strong emphasis on *differentiae* Kuhn never referred to this logical literature on taxonomies. Instead, in some of his very last writings he took interest in an Aristotelian notion of substance, and explained the no-overlap principle as based on the impossibility of the intersection of lifelines of different substances.

9. The claim that taxonomic structures are partly biologically determined comes close to views advanced by Quine as part of his naturalized epistemology. Thus, Quine also asserted that "a standard of similarity is in some sense innate. This point is not against empiricism; it is a commonplace of behavioral psychology" (Quine 1969, 11). Further, both Kuhn and Quine argued that at least parts of this biological equipment are characteristic not only of humans but of animals as well.

10. In the Postscript he also suggested that for the historian trying to identify a paradigm and its community, the circle could be broken by the new scientometric methods developed by sociologists such as Crane that would enable the identification of the community prior to investigations of its shared paradigm. Kuhn did not reflect at that time on how the identification of communities and paradigms would work for the scientists practicing science, but in his last writings from the 1990s he described a developmental solution that draws from the fact that scientists are always socialized into a particular community and then work from there (cf. Kuhn 1991).

11. In his later writings, Kuhn developed an evolutionary view in which he also ascribed the mutual isolation of different subspecialties to conceptual disparities or incommensurability: " . . . what makes these specialties distinct, what keeps them apart and leaves the ground between them as apparently empty space? To that the answer is incommensurability, a growing conceptual disparity between the tools deployed in the two specialties. Once the two specialties have grown apart, that disparity makes it impossible for the practitioners of one to communicate fully with the practitioners of the other. And those communication problems reduce . . . the likelihood that the two will produce fertile offspring" (Kuhn 2000, 120). However, the conceptual disparity between two specialties placed at different branches of the evolutionary tree of the sciences is very different from the conceptual disparity between the two specialties at each side of a revolutionary divide. The fact that there is no communication between different specialties reflects only that they address "something different", that they are not "about the same thing." On the contrary, theories such as, for example, oxygen theory and phlogiston theory are indeed "about the same thing" and therefore compete on offering the better account of their common domain (see Chang (this volume) for an analysis of the Chemical Revolution). Thus, it seems mistaken to apply the notion of incommensurability to different subspecialties.

12. For an historical case study on how differences in graded structures can lead scientists to very different assessments of anomalies, see, for example, Andersen (2009).

13. On ethnocentrism of disciplines and unidisciplinary competence, see Campbell (1969).

14. Additionally, the features used to distinguish between nodes in one taxonomy are themselves nodes in other taxonomies; however, we shall not here be concerned with this recursive element of taxonomies.
15. See Andersen (1996) for details.
16. New works in distributed cognition have started to analyze the collective cognition that happens when several people combine individual knowledge to produce a cognitive output that none of them could produce alone (Giere and Moffat 2003; Giere 2007; Kurz-Milcke, Nersessian, and Newstetter 2004; Nersessian 2004; Nersessian, Kurz-Milcke, Newstetter, and Davies 2003). However, much of this work has been concerned with agency, model systems, and the cognitive partnership between researchers and artifacts while little has yet been said about distributed conceptual structures. Likewise, work in social epistemology has started to analyze how groups arrive at joint consent to accept particular claims (Wray 1999; Wray 2006; Rolin 2008; Thagard 1997) and the role played by trust and testimony in the production of scientific knowledge (Goldman 2001; Hardwig 1991; Thagard 2005), but without analyzing how differences in conceptual structures form part of these processes and relations.

REFERENCES

Ahn, W.-K. 1998. "Why are Different Features Central for Natural Kinds and Artifacts? The Role of Causal Status in Determining Feature Centrality." *Cognition* 69: 135–178.

Andersen, H. 1996. "Categorization, Anomalies, and the Discovery of Nuclear Fission." *Studies in the History and Philosophy of Modern Physics* 27: 463–492.

Andersen, H. 2000a. "Learning by Ostension: Thomas Kuhn on Science Education." *Science & Education* 9: 91–106.

Andersen, H. 2000b. "Kuhn's Account of Family Resemblance: A Solution to the Problem of Wide-Open Texture." *Erkenntnis* 52: 313–337.

Andersen, H. 2002. "The Development of Scientific Taxonomies." In L. Magnani (ed.), *Model-Based Reasoning. Proceedings of the MBR'01*. Dordrecht: Kluwer, 95–112.

Andersen, H. 2008. "Experiments and Concepts." In U. Feest, G. Hon, H. J. Rheinberger, J. Schickore, and F. Steinle (eds.), *Generating Experimental Knowledge. Preprint 340*. Berlin: Max-Planck-Institut für Wissenschaftsgeschichte, 27–38.

Andersen, H. 2009. "Unexpected Discoveries, Graded Structures, and the Difference between Acceptance and Neglect." In J. Meheus and T. Nickles (eds.), *Models of Discovery and Creativity*. Dordrecht: Springer, 1–27.

Andersen, H., P. Barker, and X. Chen. 2006. *The Cognitive Structure of Scientific Revolutions*. Cambridge: Cambridge University Press.

Andersen, H., and N. J. Nersessian. 2000. "Nomic Concepts, Frames and Conceptual Change." *Philosophy of Science* (Proceedings) 67: 224–241.

Arabatzis, T. 2008. "Experimenting on (and with) Hidden Entities: The Inextricability of Representation and Intervention." In U. Feest, G. Hon, H. J. Rheinberger, J. Schickore, and F. Steinle (eds.), *Generating Experimental Knowledge. Preprint 340*. Berlin: Max-Planck-Institut für Wissenschaftsgeschichte, 7–18.

Barker, P. 1986. "Wittgenstein and the Authority of Science." In P. Weingartner and G. Schürz (eds.), *The Tasks of Contemporary Philosophy. Proceedings of the 10th International Wittgenstein Symposium*. Wien: Hölder-Pichler-Tempsky, 164–166.

Bower, T. G. R. 1974. *Development in Infancy.* San Francisco: Freeman.

Brown, R. 1958. "How Shall a Thing Be Called?" *Psychological Review* 65(1): 14–21.

Buchwald, J. Z. 1992. "Kinds and the Wave Theory of Light." *Studies in History and Philosophy of Science* 23: 39–74.

Campbell, D. T. 1969. "Ethnocentricism of Disciplines and the Fish-Scale Model of Omniscience." In M. Sherif and C. W. Sherif (eds.), *Interdisciplinary Relationships in the Social Sciences.* Chicago: Aldine, 328–348.

Cedarbaum, D. G. 1983. "Paradigms." *Studies in History and Philosophy of Science* 14: 173–213.

Chen, X. 1997. "Thomas Kuhn's Latest Notion of Incommensurability." *Journal for General Philosophy of Science* 28: 257–273.

Chen, X. 2002. "The 'Platforms' for Comparing Incommensurable Taxonomies: A Cognitive-Historical Analysis." *Journal for General Philosophy of Science* 33: 1–22.

Chi, M. T. H., J. Feltovich, and R. Glaser. 1981. "Categorization and Representation of Physics Problems by Experts and Novices." *Cognitive Science* 5: 121–152.

Duschl, R. A. 1991. "Epistemological Perspectives on Conceptual Change: Implications for Educational Practice." *Journal of Research in Science Teaching* 28: 839–858.

Giere, R. N. 1994. "The Cognitive Structure of Scientific Theories." *Philosophy of Science* 61: 276–296.

Giere, R. N. 2007. "Distributed Cognition without Distributed Knowledge." *Social Epistemology* 21: 313–320.

Giere, R. N., and B. Moffat. 2003. "Distributed Cognition: Where the Cognitive and the Social Merge." *Social Studies of Science* 33: 301–310.

Goldman, A. I. 2001. "Experts: Which Ones Should You Trust?" *Philosophy and Phenomenological Research* 63: 85–110.

Gooding, D. 1986. "How Do Scientists Reach Agreement about Novel Observations?" *Studies in History and Philosophy of Science* 17: 205–230.

Hardwig, J. 1991. "The Role of Trust in Knowledge." *Journal of Philosophy* 88: 693–708.

Hoyningen-Huene, P. 1993. *Constructing Scientific Revolutions: Thomas S. Kuhn's Philosophy of Science.* Chicago: University of Chicago Press.

Keil, F. 1979. *Semantic and Conceptual Development.* Cambridge, MA: Harvard University Press.

Kindi, V. 1995. "Kuhn's *The Structure of Scientific Revolutions* Revisited." *Journal for General Philosophy of Science* 26: 75–92.

Kuhn, T. S. 1959. "The Essential Tension: Tradition and Innovation in Scientific Research." In C. W. Taylor and F. Barron (eds.), *Scientific Creativity: Its Recognition and Development.* New York: John Wiley & Sons, 341–354.

Kuhn, T. S. 1970a. *The Structure of Scientific Revolutions.* Chicago: The University of Chicago Press.

Kuhn, T. S. 1970b. "Logic of Discovery or Psychology of Research?" In I. Lakatos and A. Musgrave (eds.), *Criticism and the Growth of Knowledge.* Cambridge: Cambridge University Press, 1–24.

Kuhn, T. S. 1974. "Second Thoughts on Paradigms." In F. Suppe (ed.), *The Structure of Scientific Theories.* Urbana: University of Illinois Press, 459–482. Reprinted in T. S. Kuhn, *The Essential Tension.* Chicago: The University of Chicago Press, 293–319.

Kuhn, T. S. 1979. "Metaphor in Science." In A. Ortony (ed.), *Metaphor in Science.* Cambridge: Cambridge University Press, 409–419.

Kuhn, T. S. 1983a. "Commensurability, Comparability, Communicability." *PSA* 2: 669–688.

Kuhn, T. S. 1983b. "Response to Commentaries" [on Commensurability, Comparability, Communicability]. *PSA* 2: 712–716.

Kuhn, T. S. 1989. "Speaker's Reply." In S. Allén (ed.), *Possible Worlds in Humanities, Arts, and Sciences.* Berlin: de Gruyter, 49–51.

Kuhn, T. S. 1991. "The Road Since Structure." *PSA 1990* 2: 3–13.

Kuhn, T. S. 1993. "Afterwords." In P. Horwich (ed.), *World Changes.* Cambridge, MA: The MIT Press, 311–341.

Kuhn, T. S. 2000. "The Trouble with the Historical Philosophy of Science." In T. S. Kuhn, *The Road Since Structure.* Edited by J. Conant and J. Haugeland. Chicago: University of Chicago Press, 105–120.

Kurz-Milcke, E., N. Nersessian, and W. Newstetter. 2004. "What Has History to Do with Cognition? Interactive Methods for Studying Research Laboratories." *Culture and Cognition* 4: 663–700.

Laurence, S., and E. Margolis. 1999. "Concepts and Cognitive Science." In E. Margolis and S. Laurence (eds.), *Concepts: Core Readings.* Cambridge MA: MIT Press, 3–81.

Linn, M. C. 2008. "Teaching for Conceptual Change: Distinguish or Extinguish Ideas." In S. Vosniadou (ed.), *International Handbook of Research on Conceptual Change.* New York: Routledge, 694–722.

Markman, E. M. 1989. *Categorization and Naming in Children: Problems of Induction.* Cambridge, MA: MIT Press.

Murphy, G., and D. Medin. 1999. "The Role of Theories in Conceptual Coherence." In E. Margolis and S. Laurence (eds.), *Concepts.* Cambridge, MA: MIT Press, 425–458.

Nersessian, N. J. 1992. "Constructing and Instructing: The Role of 'Abstraction Techniques' in Creating and Learning Physics." In R. A. Duschl and R. J. Hamilton (eds.), *Philosophy of Science, Cognition, Psychology and Educational Theory and Practice.* Albany: SUNY Press, 48–53.

Nersessian, N. J. 1995. "Should Physicists Preach What They Practice? Constructive Modeling in Doing and Learning Physics." *Science & Education* 4: 203–226.

Nersessian, N. 1998. "Conceptual Change." In W. Bechtel and G. Graham (eds.), *A Companion to Cognitive Science.* Oxford: Blackwell, 157–166.

Nersessian, N. J. 2003. "Kuhn, Conceptual Change, and Cognitive Science." In T. Nickles (ed.), *Thomas Kuhn.* Cambridge: Cambridge University Press, 178–211.

Nersessian, N. 2004. "Interpreting Scientific and Engineering Practices: Integrating the Cognitive, Social and Cultural Dimension." In M. Gorman, R. Tweney, D. Gooding, and A. Kincannon (eds.), *Scientific and Technological Thinking.* Mahwah, NJ: Erlbaum, 17–56.

Nersessian, N. J. 2008a. *Creating Scientific Concepts.* Cambridge, MA: MIT Press.

Nersessian, N. J. 2008b. "Mental Modelling in Conceptual Change." In S. Vosniadou (ed.), *International Handbook of Research on Conceptual Change.* New York: Routledge, 391–416.

Nersessian, N., E. Kurz-Milcke, W. Newstetter, and J. Davies. 2003. "Research Laboratories as Evolving Distributed Cognitive Systems." In R. Alterman and D. Kirsch (eds.), *Proceedings of the Twenty-Fifth Annual Conference of the Cognitive Science Society,* 857–862

Quine, W. V. O. 1969. "Natural Kinds." In N. Rescher (ed.), *Essays in the Honor of Carl G. Hempel.* Dordrecht: Kluwer, 5–23.

Rolin, K. 2008. "Science as Collective Knowledge." *Cognitive Systems Research* 9: 115–124.

Rosch, E. 1978. "Principles of Categorization." In E. Rosch and B. B. Lloyd (eds.), *Cognition and Categorization.* Hillsdale, NJ: Erlbaum, 27–48.

Rosch, E., C. B. Mervis, W. D. Gray, D. M. Johnson, and P. Boyes-Braem. 1976. "Basic Objects in Natural Categories." *Cognitive Psychology* 8: 382–439.

Sloman, S. A., B. C. Love, and W. K. Ahn. 1998. "Feature Centrality and Conceptual Coherence." *Cognitive Science* 22: 189–227.

Thagard, P. 1992. *Conceptual Revolutions*. Princeton: Princeton University Press.

Thagard, P. 1997. "Collaborative Knowledge." *Nous* 31: 242–261.

Thagard, P. 2005. "Testimony, Credibility, and Explanatory Coherence." *Erkenntnis* 63: 295–316.

Treagust, D. F., and R. Duit. 2008. "Conceptual Change: A Discussion of Theoretical, Methodological and Practical Challenges for Science Education." *Cultural Studies of Science Education* 3: 297–328.

Vosniadou, S. 2007. "The Cognitive-Situative Divide and the Problem of Conceptual Change." *Educational Psychologist* 42: 55–66.

Wray, K. B. 2006. "Scientific Authorship in the Age of Collaborative research." *Studies in History and Philosophy of Science* 37: 505–514.

Wray, K. B. 1999. "A Defense of Longino's Social Epistemology." *Philosophy of Science* 66: S538–S552.

Additional Archival Material

Letter from Thomas S. Kuhn to Eleanor Rosch, 7 December 1956. Thomas S. Kuhn Papers, MC 240, box 12, folder 4. Massachusetts Institute of Technology, Institute Archives and Special Collections, Cambridge, Massachusetts.

Thomas S. Kuhn: Philosophy 24.853. Thomas S. Kuhn Papers, MC 240, box 23, folder 5. Massachusetts Institute of Technology, Institute Archives and Special Collections, Cambridge, Massachusetts.

Thomas S. Kuhn: "An Historian's Theory of Meaning." Talk to Cognitive Science Colloquium, UCLA, 26 April 1990. Thomas S. Kuhn Papers, MC 240, box 24, folder 8. Massachusetts Institute of Technology, Institute Archives and Special Collections, Cambridge, Massachusetts.

10 Kuhn, Naturalism, and the Social Study of Science

Alexander Bird

1. NATURALISM IN THE WORK OF THOMAS KUHN

One of the most interesting features of Thomas Kuhn's work in *The Structure of Scientific Revolutions* (SSR) is its naturalism. From a philosophical point of view, this naturalism was highly original, sufficiently original that it was not properly understood, either by critics or by many supporters of Kuhn's work. Quine (1969) famously proposed methodological naturalism as an approach to epistemology, seven years after the publication of the first edition of SSR, and only just before the publication of the second edition, in which a postscript clarified some of the more important naturalistic elements of the book. It is no wonder that Kuhn said of his notion of *exemplar* (the central application of the concept *paradigm*), that this is the "most novel and least understood aspect of [SSR]" (Kuhn 1970, 187).

There are two strands to Kuhn's naturalism. The first concerns his use of history. SSR contains a lot of history. But it is not a historical work in a straightforward way (unlike his 1978 *Black-Body Theory and the Quantum Discontinuity, 1894–1912*). Rather, Kuhn uses historical examples to generate a general account of the (structure of the) historical development of science, an account that has a significant social element. This in turn he puts to philosophical use, fulfilling his original motivation for this work. Philosophers of science had used historical examples before, but not in this way. Popper uses historical examples, but the philosophical purposes are strictly limited. Popper's work is normative; the use of examples is to show that science does indeed live up to these norms—a question that he would have regarded as not strictly philosophical at all. Nor is Popper's use of historical examples systematic; in common with other philosophers of science, Popper's use is illustrative rather than marshaled as evidence. Kuhn instead makes a powerful case for a pattern in the history of science, an alternation of normal and revolutionary periods of science. This contrasts with the cumulative account, whereby new well-confirmed beliefs are added to the preexisting stock of well-confirmed beliefs. The cumulative account is what one would expect (perhaps subject to some idealizing assumptions) if standard accounts of theory confirmation (as one may find among the positivists) were correct.

The second strand of Kuhn's naturalism concerns his willingness to deploy evidence drawn from psychology. For example he cites the experiments of the psychologists Bruner and Postman as evidence for a thesis concerning the theory-ladenness of observation (1962, 62–63). He also refers to the work of the Gestalt psychologists in articulating his thesis of world change—how the world appears differently after a revolution (e.g., 1962, 2, 111–114). He also briefly mentions the work of Jean Piaget and of Benjamin Whorf (1962, 2). Kuhn (1970, 191–192) even remarks that he was experimenting with a computer program to model shared intuitions about similarity. While he uses gestalt images primarily as an analogy, Kuhn does raise the possibility that there is an underlying unity to both Gestalt psychology and to the cognitive processes involved in revolutionary scientific change, and refers to other psychological work as going in this direction, especially that of the Hanover Institute; he footnotes works by the psychologists Stratton and Carr concerning accommodation to image-inverting glasses. That the psychology of gestalt images and the psychology of paradigm change are linked is an entirely reasonable speculation for Kuhn. Kuhn's most significant idea in SSR is the idea of an exemplar. This is the core of the concept of paradigm; an exemplar is an exemplary, paradigmatic puzzle solution that is used in the training of scientists. This training helps establish *similarity relations* for the scientist. Similarity relations are dispositions to see one thing as like another; in particular these enable the scientist to see one puzzle as similar to another puzzle and, likewise, one puzzle solution as similar to another solution. A scientific revolution involves the introduction of new puzzle solutions to solve particularly significant, anomalous puzzles. These new solutions become the new exemplars. Consequently new similarity relations are set up and old ones are broken down. Since such similarity relations structure how the subject apprehends the world, scientific revolutions cause a change in the nature of the subject's apprehending. "Apprehend," as I have used it, can mean the visual appearance of the world, and this would apply to cases similar to the Bruner and Postman experiments and those discussed by the Gestalt psychologists. "Apprehend" may also mean the subject's understanding of the more abstract structure or pattern of ideas in a scientific theory.

The two naturalistic strands (historical-sociological and cognitive-psychological) are not independent. For exemplars, by setting similarity relations for puzzles and their solutions, are the motor of normal science. Kuhn (1970, 189–190) tells us, "The role of acquired similarity relations also shows clearly in the history of science. Scientists solve puzzles by modeling them on previous puzzle-solutions, often with only minimal recourse to symbolic generalizations. Galileo found that a ball rolling down an incline acquires just enough velocity to return it to the same vertical height on a second incline of any slope, and he learned to see that experimental situation as like the pendulum with a point-mass for a bob." And it is the failure of anomalous puzzles to find a solution by reference to these exemplars that gives rise to crisis and so to revolution. Consequently the cognitive role of exemplars explains the historical cycle of normal science and revolution. This is important, because it adds to the philosophical significance of Kuhn's historical work. This point is best appreciated by comparison to the role of history in Popper's philosophy. Above, I suggested that historical examples are subsidiary to Popper's principal normative task. Popper's examples are intended to show that as a matter of fact scientists do obey the norms he sets out. But it would not immediately invalidate the normative claims to discover that scientists do not always obey the norms—one would just have to conclude that scientists are not always as rational as one might have hoped. A similar response might have been made on behalf of the positivists in response to Kuhn's historical studies if all Kuhn had shown was that sometimes science undergoes noncumulative revolutions. That is prima facie consistent with the claim that had scientists been perfectly rational, such episodes would not have occurred. The fact the Kuhn offers a well-evidenced *pattern* puts some pressure on such a reply: how could individual cases of irrationality explain a pattern in science? More importantly, Kuhn's explanation in terms of exemplars provides an alternative explanation: crises and revolutions are not the outcome of localized irrationality, but rather of deep facts about human (individual and social) cognition. Kuhn is offering not an alternative to the rationality of science but an alternative account of what rationality consists in; it is not a matter of following some a priori rule of method but is rather a matter of using acquired similarity relations to match puzzles and solutions. I shall return to Kuhn's conception of rationality below.

As it was, the significance of Kuhn's ideas was not properly appreciated at the time. That was partly because Kuhn's broad use of the term "paradigm" obscured the core idea of the paradigm-as-exemplar, requiring him to make this idea more explicit in the Postscript to the second edition of SSR. It was also because Kuhn's naturalism, in both respects, was novel in philosophy. As mentioned, Kuhn's naturalism predated Quine's "Epistemology Naturalized" by the best part of a decade. And more importantly, Quine was only proposing naturalism, not exhibiting it—it was Kuhn's naturalistic *practice* (without a wrapping of philosophical explanation) that nonplussed philosophers. A third aspect was social. The fact that science is a social enterprise links the cognitive function of exemplars in individuals to the large-scale pattern in the history of science. So individual scientists, during normal science at least, acquire their similarity relations (their ability to recognize similarities between puzzles and between puzzle solutions, etc.) by enculturation during the process of education and training. This key social element in Kuhn's story contrasted starkly with the individualist paradigm in epistemology and so seemed inconsistent with the possibility of justified belief in science. His view being misunderstood, Kuhn was taken to endorse an irrationalist conception of science: according to this accusation, he held that the development of science is the manifestation of "mob psychology" (Lakatos 1970, 178).

While this meant that Kuhn's work was underappreciated by philosophers of science, it did also lead to its being held to provide an opportunity for other students of science. Not only does the Kuhnian picture give a central role to social practices (in enculturating paradigms), it permits a wider integration of science as a field of activity into the traditional concerns of historiography and sociology. For if the outcome of a scientific dispute is not determined by rules of rationality, then there is room for those outcomes being determined by social and political forces, just as they are outside science. Historians and social theorists had earlier tended to think that even if there is some room for explaining the social setting of science and how this influences the general development of science, there was only limited opportunity for distinctively historical and social explanations of the content of the particular theories that prevailed. So while Robert K. Merton (1938) linked the scientific revolution in seventeenth-century England to Protestant pietism, that explanation did not seek to explain the detailed products of that science, such as Harvey's hypothesis of the circulation of the blood, Newton's gravitational theory, or Boyle's law. Kuhn's approach seemed to open up the possibility that these too would submit to historical and social explanation—Kuhn's work lies at the origin of the contemporary study of science, inaugurating what Collins and Evans (2002) call the "second wave" in science studies. The most significant and philosophically most sophisticated school within this second wave is the Edinburgh School's Strong Programme in the Sociology of Scientific Knowledge (SSK). The latter was formulated as such in explicit contrast to the *weak* programme, which desists from applying sociological analysis to the content of scientific belief except in pathological cases.

Kuhn (1992, 8–9) himself, in the strongest terms, repudiated the Strong Programme, which he claimed

> has been widely understood as claiming that power and interest are all there are. Nature itself, whatever that may be, has seemed to have no part in the development of beliefs about it. Talk of evidence, of the rationality of claims drawn from it, and of the truth or probability of those claims has been seen as simply the rhetoric behind which the victorious party cloaks its power. What passes for scientific knowledge becomes, then, simply the belief of the winners.
>
> I am among those who have found the claims of the strong program absurd: an example of deconstruction gone mad. And the more qualified sociological and historical formulations that currently strive to replace it are, in my view, scarcely more satisfactory.

While it may rightly be said that Kuhn is responding to a caricature of the Strong Programme, these remarks nevertheless demonstrate that there is a significant gulf between Kuhn and those who regard themselves as working in a Kuhnian tradition. Donald MacKenzie (1981, 2–3) contrasts two views of the sociohistorical investigation of science, corresponding to the weak programme (or first wave of science studies) and to the Strong Programme (the second wave). The former discusses "factors [that] could hinder or promote work in the field and perhaps condition the quality of the work done," but which "are, however, not taken . . . as explaining the content of the theoretical advances." The latter is exemplified by Joseph Ben-David's work (1971) on the growth of statistics in the United States and Britain and the social context that allowed particularly rapid theoretical advance in Britain. MacKenzie regards his own work, which links the content of these advances to the promotion of eugenics in the interests of Britain's professional middle classes, as exemplifying the second view, according to which social factors play a role in the process of producing new ideas, and, furthermore (in a stronger version of the second view), in determining their acceptance or rejection. SSR, says MacKenzie (1981, 3), "can be read as a statement of this second view in its strong version."

At this point it will be useful to make two distinctions that are both vague and contested. For our purposes, the more important is the distinction between explanatory factors that are internal to science and those that are external to science. The rough idea of this difference is clear: if a scientist promotes adoption of an idea because she believes it solves an agreed puzzle, then that explanation is internal to science; but the explanation is external if, for example, the idea is promoted because the scientist, who holds a patent, has a financial stake in its being believed. The key distinction may be taken to be something like this: a factor that influences theory construction and choice is internal to science if it is among those factors that the scientific community takes to be rational in the sense of being conducive to the epistemic aims of science. Otherwise a factor is external. When Copernicus retained the circular motion central to Ptolemy's cosmology, he did so because he reasonably believed this to be a feature that would contribute to accurate puzzle solving in cosmology, and so it is an internal feature. By contrast, according to Farley and Geison (1974), what they call "extrinsic" factors such as conservative politics and religion played a part in forming Pasteur's views and enabling their triumph over those of Pouchet; the amenability of a scientific idea to a certain religious or political perspective does not lead to improved puzzle solving, nor would it reasonably have appeared so to the scientific community of the time, and so such factors (if they did play this role) are external.

Whether or not one regards this distinction as important or useful (or indeed one that can be properly made at all), Kuhn himself employed it. As MacKenzie correctly notes, "In Kuhn's work the society that influences scientific judgment is typically taken to be the community of practising scientists in a particular specialty." For Kuhn those factors are therefore internal.

MacKenzie himself (1981, 4) regards the factors he discusses as including external factors, and he sees this as an extension of Kuhn's approach:

> But there seem to be no overwhelming grounds why only social factors internal to the community of scientists must necessarily be at work. If the basic point of the social nature of scientific judgement is taken, then it seems reasonable to search society at large, as well as the scientific community, for determinants of it.

MacKenzie clearly sees his externalist approach as being a natural extension of the lessons learned from Kuhn. As I shall explain, this is gravely to misunderstand Kuhn. As we have seen, Kuhn distances himself from the Strong Programme, and even if the strength of his rhetoric is not justified by a more nuanced understanding of the Strong Programme's aims, it nonetheless remains the case that Kuhn had excellent reasons for maintaining a robustly internalist approach, one that is very much closer to Ben-David than to MacKenzie.

The second distinction is between microsocial and macrosocial explanations. Macrosocial explanations refer to large-scale social structures and changes, such as political, religious, or economic forces, that play a role in many aspects of society. Microsocial explanations focus on interactions between agents in fairly restricted groups. A macrosocial investigation of industrial unrest would be concerned with the economic climate, social attitudes to strikes, relationship with other political agendas, legislative context, and the like. A microsociological study would focus on the various kinds of relationship of power and interest (and perhaps collegiality or suspicion) at work between union officials, industrial bosses, the workers, and party leaders. The two kinds of study are not mutually exclusive and may overlap, but the kinds of focus are clearly different. The first wave of science studies concerned itself with macrosociological explanations. Merton's work is a notable example, as is Hessen's "The Social and Economic Roots of Newton's *Principia*" (1971). Microsocial studies are more recent, exemplified by the work of Collins (2004) on the interactions of networks of individual scientists working on gravity waves.

The relationship between externalism/internalism and macro/micro is complex. Whether a macrosociological study is externalist or not will depend on what it aims to explain. If it aims to explain the content of the beliefs that come to be accepted as knowledge, then it will be externalist. Merton does not aim to tell us why *these* theories were adopted in the scientific revolution and not those (neither, for that matter, does Hessen). MacKenzie, on the other hand, does aim at showing how the content of statistical theory was influenced by eugenics and the interests of the professional class, and so is externalist. When it comes to microsocial explanations, the distinction between internalist and externalist explanation is less straightforward (and in many cases less pertinent). Typical of such explanations is a reference to the interests of an actor. If such explanations are epistemic, then the explanation is straightforwardly internalist. But even interests that are not epistemic, such as professional ambition, might be consistent with having other motivations that are. Thus the microsocial explanations in Collins' work can be read in a mostly internalist way.

So, to conclude, we have two strands to Kuhn's naturalism: a historical-sociological one, which led to a major trend in science studies, a trend which Kuhn himself rejected. And we have a cognitive-psychological naturalism that was rejected by Kuhn's philosophical contemporaries. While the latter has been underappreciated, the former has been overappreciated. Neither has been properly understood. I shall concentrate largely on the science studies movement as putative heirs to Kuhn's sociohistorical naturalism before returning to discuss the recent rehabilitation of Kuhn's cognitive-psychological naturalism. In this article I argue that insofar as they take a predominantly externalist approach, students of science studies are not Kuhn's true heirs. Kuhn had very good theoretical reasons for rejecting externalism. I shall explain why Kuhn did, unwittingly, play a major role in creating the environment in which science studies could flourish. After suggesting that Durkheim's functionalism would provide a suitable basis for future sociology of science, I shall conclude by suggesting that the naturalistic study of science has hitherto been best represented by recent work on the psychology of scientific research, in particular work on model-based and analogical reasoning.

2. KUHN AND EXTERNALISM IN THE STUDY OF SCIENCE

In a phrase that has elicited scorn from Steve Fuller (2002), I have described SSR as "theoretical history of science" (2000, 29), by which I mean it has two components that have analogues in natural science:

(i) a *descriptive* element—SSR identifies a general pattern in the development of science: science is a puzzle-solving enterprise which shows a cyclical pattern of normal science, crisis, revolution, normal science . . . ;

(ii) an *explanatory* element—it proposes an explanation of the pattern identified in (i): puzzle solving is driven by adherence to a *paradigm* (an exemplary puzzle solution).

It is important to appreciate that if Kuhn's theory is to be correct, the drivers of scientific change must be largely internal to science itself. Just taking the descriptive element on its own, it would be odd if the development of science showed this regular, cyclical pattern yet the causes of the pattern were to be found outside science. One might reconcile patterns in science with external forces, if one could show that those external forces themselves show a similar kind of regularity. But that is implausible. The oddity in my phrase "theoretical history" is precisely that historians for the most part do not believe that there are patterns in history; and even those that have been proposed, such as Marx's, do not fit with Kuhn's patterns. Kuhn opposed the extension of his ideas beyond science—so a true Kuhnian could not claim that the pattern found in science is an instance of some more general pattern found in the development of, for example, societies at large. If there is a Kuhnian pattern in science, then the best explanation of it will be one that refers to factors internal to the activity of science. And indeed, as we know, that is precisely what Kuhn offers us in the explanatory part of his theory. During normal science scientific advance is driven by a paradigm or set of paradigms, which are exemplary scientific puzzle solutions. Kuhn spells out in detail how these fulfill a myriad key functions in normal science: fixing concepts, training scientists to see puzzle solutions, providing a standard for the assessment of proposed puzzle solutions, and so forth. Since, Kuhn emphasizes, the bulk of scientific activity is normal science, it follows that at least most scientific change is governed by factors internal to science.

Consequently, Kuhn has good reasons stemming from the central commitments of his theory for maintaining a predominantly internalist picture of science: " . . . the ambient intellectual milieu reacts on the theoretical structure of a science only to the extent that it can be made relevant to the concrete technical problems with which the practitioners of the field engage" (1971, 137–138). Kuhn does make room for two ways in which external forces do play a part in science. First, such factors may be significant at the inception and early development of a discipline (Kuhn 1968, 118). Second, they may play a role in determining the timing of a scientific development (Kuhn 1962, 69; 1968, 118). Note that neither of these contradicts the position articulated above, of which the central claim is that the content of scientific change is determined by factors internal to a tradition. Tautologously, that tradition cannot be brought into existence by factors internal to it. And it is no rejection of the claim to accept that the pace of change is constrained by external factors such as the resources society is able to put into science. These exceptions thus prove the Kuhnian rule, that scientific change is driven by the internal requirement of problem solving. Returning to MacKenzie's contrast between Ben-David's view of social forces and his own, we see that Kuhn is much closer to Ben-David than to MacKenzie. Both Kuhn and Ben-David restrict the principal effects of social causes on the development of a field to its acceleration or retardation and reject effects on change in its content. Hence it is at best misleading to portray the extension of science studies to include social causes of change in content as a further step in a direction Kuhn had already pointed us in. Whatever may be true of students of science in the "first wave," adherence to internalism is not a "failure of nerve" (Bloor 1991, 4) in Kuhn's case, but a theoretically well-motivated commitment.

A natural response may be to point to revolutionary science as being the locus of the play of external social forces within a Kuhnian framework. Yet I believe this shows a misunderstanding of Kuhnian revolutions, and that as regards the principal driver of scientific progress, there is considerable continuity between normal and revolutionary science; for in both the central concern is with the need to solve scientific puzzles. What differs is the manner in which this driver is manifest. During normal science, the need to solve problems is satisfied by the paradigm. During extraordinary science the need remains, but now must be satisfied by finding a replacement paradigm. What determines the outcome will still be, principally, the power of a proposed paradigm to solve puzzles. That may not determine the outcome uniquely—Kuhn stresses that there is room for rational disagreement about the relative problem-solving power of a proposed new paradigm compared to the old one or a competitor. Nonetheless, the fact that the dispute is about scientific puzzle-solving power puts significant constraints on the choices available. The participants in the debate must be able rationally to believe that their favored solution will deliver more and better puzzle solutions than its competitors. In particular supporters of a new paradigm must, in most cases, be able to show that it solves a sizeable proportion of the most significant anomalies that beset the old paradigm, while also preserving the bulk of the puzzle-solving power of its predecessor. Since finding an innovative solution that achieves this is not easy, most episodes in revolutionary science will provide very few choices, typically just the old paradigm and a single revolutionary proposal. Many critics, from Toulmin (1970) onwards, myself included, have argued that Kuhn overemphasizes the difference between normal and revolutionary science. But we should not ourselves read into Kuhn an account of the difference that makes the two utterly unlike one another; Nickles (2012), in this volume, suggests that the best Kuhnian account of the use and development of exemplars should allow for innovation with regards to exemplars in normal science as well as the retention of exemplars through extraordinary science. There is a caricature of Kuhn's scientific revolutions as root-and-branch revisions, complete breaks with the past, whose outcomes are determined in a manner quite unlike the determination of episodes in normal science. Yet Kuhn himself argued for significant similarities between the two kinds of episode. The final chapters of *The Structure of Scientific Revolutions*, "The Resolution of Revolutions" and "Progress through Revolutions" describe the constraints imposed on the new paradigm by the long-standing success in puzzle solving of its predecessor. Such constraints, later amplified by an articulation of five universal scientific values (accuracy, consistency, breadth of scope, simplicity, fruitfulness), mean that there is significant continuity in revolutionary science (see Kuhn 1977). Consequently, given the infinite range of beliefs a scientist could have about a given subject matter, all but a small handful are excluded by factors internal to science, even during extraordinary science.

The moral of this section is that consideration of Kuhn's mode of explanation for both normal and revolutionary science is incompatible with significantly externalist elements in an explanation of a scientific development. Given that Kuhn states his internalism and that his theory gives him reason to do so, why has Kuhn been seen as a promoter of externalism? In particular why has the (internalism-supporting) continuity afforded by the unchanging puzzle-solving imperative in Kuhn's picture not been sufficiently appreciated? I suggest that there are two reasons. The first is the overemphasis given to one unrepresentative passage in SSR. The second centers on the alleged irrationalism of Kuhn's view, which if it were correct, would fit well with a significant role for social influence on the content of science.

One passage in SSR has been a particular source for those who wish to take an externalist lesson from Kuhn:

> Individual scientists embrace a new paradigm for all sorts of reasons and usually for several at once. Some of these reasons—for example, the sun worship that helped make Kepler a Copernican—lie outside the apparent sphere of science altogether. Others may depend upon idiosyncrasies of autobiography and personality. Even the nationality or the prior reputation of the innovator and his teachers can sometimes play a significant role.

On inspection, this passage (Kuhn 1962, 152–153) is not so thoroughly externalist as may be supposed. On the one hand Kepler's sun worship is certainly externalist, and so is nationality. But the other factors are not. Note that Kuhn *contrasts* these other factors with the clearly externalist sun worship. So Kuhn (1977, 325) tells us that "some scientists place more premium than others on originality and are correspondingly more willing to take risks," for example. Frank Sulloway (1996) has shown that birth order plays a significant part in one's readiness to accept unorthodox opinions. Such a readiness, a rebellious temperament, is part of one's personality. Should this count as externalist? Arguably yes, since it is a cause of behavior that originates outside science. Yet, at the same time what is affected is a differential response to the relationship between evidence and theory. Below I shall argue that Kuhn has a conception of scientific rationality that accommodates disagreement about the import of evidence. Indeed, Kuhn himself suggests elsewhere that such individual differences are healthy for scientific progress, since a greater variety of ideas will be pursued, and that the mean effect of such differences coincides with what a more traditional view would regard as rationally recommended. Also under the heading of biography are factors that "result from the individual's experience as a scientist. In what part of the field was he at work when confronted by the need to choose [between theories]? How long had he worked there; how successful had he been; and how much of his work depended on concepts and techniques challenged by the new theory?" (Kuhn 1977, 325). While these are matters of individual biography, they relate, as Kuhn says, to the individual's experience *as a scientist*. Kuhn also mentions reputation, a social category. As I noted above social explanations are not necessarily externalist. If a scientist's opinion carries weight because she has an impressive scientific record, then that factor is unproblematically internalist; it remains the case that the community's driving interest is effective puzzle solving.

More importantly, not only is this short passage only partly externalist, it is followed by a much longer and more detailed discussion of clearly internalist factors, which commences, "Probably the single most prevalent claim advanced by the proponents of a new paradigm is that they can solve the problems that have led the old one to a crisis. When it can legitimately be made, this claim is often the most effective one possible" (Kuhn 1962, 153). And it is most effective when "the new paradigm displays a quantitative precision strikingly better than its older competitor" (1962, 153–154). Kuhn then goes on to point out that solving the old paradigm's anomalies is not always necessary, for sometimes the new paradigm candidate might not provide solutions to the crisis-evoking problems. In that case novel predictions, predictions of phenomena that would be entirely unexpected under the old paradigm, can be persuasive (such as the prediction of the phases of Venus by Copernicus' theory). Kuhn then mentions the role of aesthetic considerations. He also discusses at length the characteristic of revolutions we have subsequently called "Kuhn loss" and the importance of a new paradigm being a fruitful basis for new problem-solving research. In assessing whether Kuhn gave direct encouragement to externalist study of science, we must set the short quoted passage against the six pages that follow.

Those who promote SSK often draw upon the Duhem–Quine thesis to support their view that social and political factors must play a role in determining which scientific theories are adopted.[1] The idea is that rationality plus the evidence is never enough to determine a single scientific conclusion—there are always equally acceptable competing hypotheses available. Consequently, factors in addition to the evidence must always play a part in explaining the choices that are actually made, and one natural supposition is that these factors are social and political. At first sight, Kuhn's picture seems to fit neatly into this pattern. Does he not avow a sociological conception of "paradigm" which he renames the "disciplinary matrix," emphasizing the socialization of new recruits to the scientific profession?

While that is true, it does not support the thesis that external elements play a determining role. During normal science successful puzzle solutions are usually determined by the evidence plus the background scientific commitments of the scientist. Take for example the normal science task of determining with greater accuracy the value of some important constant. Frequently, so long as the task is carried out competently, there will be no room for alternatives. The fact that the background commitments employed in this process may be acquired by a process of socialization into a disciplinary matrix does not mean that they are extrascientific. As I have just argued, even in Kuhn's account of scientific revolutions the factors that influence decisions are pre-dominantly internal to science. It may be true that such factors do not fix a *unique* rational preference—Kuhn tells us that in such periods rational men may disagree.[2] But this is not to concede that there is need to appeal to exter-nal factors. Consider two men watching a sporting contest, who disagree at halftime about who will win: Dave maintains that Bath will win because they have scored more points in the first half while Mick maintains that Bristol's greater ball possession and increasing dominance will lead to their victory. They may agree on the evidence but weigh its significance differently. It need not be that their difference in opinion is motivated by considerations external to the game in hand. A notion of rationality is needed that does not mandate (at most) a single rationally acceptable outcome among competing theories. Traditional accounts of scientific method do have the consequence that there is at most a single rational response to the evidence, whereas Kuhn *rejects* such a view.

The underdetermination arguments of Barnes et al. employ a conception of rationality inherited from logical empiricism.[3] The latter tends to see ratio-nal scientific inference as an extension of deductive logic (Carnap's inductive logic is the prime example). If we employ such a conception of rationality it will indeed appear, as Barnes (1990, 63) tells us, that "reason and experience do not sufficiently determine the knowledge which an individual scientist should adopt." However, the extent to which reason and experience fail to determine scientific choices rather depends on what counts as "reason" (and for that matter, "experience" also). Since a richer conception of reason will make a choice more determinate whereas a thin conception of reason—for example, one limited to deductive logic—will leave a great deal more unde-termined, Barnes' claim would appear to depend on his accepting the limited account of reason just mentioned. That commitment, it may be noted, would thereby seem to be in tension with the rationality version of the symmetry principle, as presented by Barnes and Bloor (1982). For the latter should pre-clude the analyst from adopting any specific account of what is rational or not. A more Kuhnian conception of rationality is not only less susceptible to the underdetermination argument; it will also be, I think, consistent with the relativism implied by the symmetry principle. For as our acquired similarity relations change, so will what it is rational to believe, even if the evidence per se does not change. Furthermore, unlike the logical empiricist account, it will allow that incompatible propositions might be believed by different, rational individuals on the basis of the same evidence. Although one might wish to explain differences in opinion, if differing conclusions may both be rational responses to the evidence, acceptable explanations of that difference need not include external social explanations.

This leads to the second reason why externalism has been attributed to Kuhn: his alleged irrationalism. The standard logical empiricist account says that scientific change is answerable to (if not exactly driven by) unchanging rational rules of logic and confirmation; these rules, which may be applied algorithmically, determine a unique rational answer. Kuhn (1970, 199) remarks, "Debates over theory-choice cannot be cast in a form that fully resembles logical or mathematical proof." While Kuhn thinks this point has long been familiar in philosophy of science, his own articulation of the process of theory choice was not familiar: "The practice of normal science depends on the ability, acquired from exemplars, to group objects and situations into similarity sets which are primitive . . ." (1970, 200). The divergence between the logical empiricist account of scientific thought and Kuhn's meant that the latter was immediately characterized by his detractors as an irrationalist view, as Kuhn himself (1970, 199) notes. But once one has decided that Kuhn is portraying scientists as irrational it doesn't much matter where in his picture the influences on their choices come from. As Roy Porter (1990, 38) says,

> Because Kuhn argued that paradigms were strictly speaking incommensurable, i.e., one could not be judged better than its predecessor according to an accessible objective truth criterion, his opponents accused him of encouraging a subjectivist and relativist view of the status of science. If that were so, factors which were (in the wider sense) ultimately social would count, after all, in theory choice. Kuhn later backtracked on various of the inferences derived from his ideas, but his hyper-influential book gave impetus and respectability to many new currents from the mid-1960s.

The difference between internal and external causes of irrational belief hardly mattered and was overlooked by his critics as insignificant when compared to the important distinction between rational and irrational belief. I suspect that many in the science studies world overlooked the distinction for the same reason, although taking a different view of whether the charge amounted to a criticism of Kuhn.[4]

Kuhn spends much of the Postscript resisting the charges of subjectivism and irrationalism. This contrasts with his response to the charge of relativism, which, with appropriate explanations, he is willing to accept (1970, 207). We should conclude therefore that Kuhn is not rejecting rationalism about science.[5] He rejects a particular, logical empiricist account of rationality, one that sees it as an extension of deductive logic and which denies the possibility of rational disagreement. And in place of the latter, Kuhn is in effect offering a different conception of what rational science is: one that is anchored in certain shared similarity relations.

There are two features to the shared similarity relations that are important for understanding Kuhn's revisionary approach to rationality. The first is social: these are *shared* relations. For example, Kuhn tells us that theory choice is dependent upon the values scientists hold (such as consistency, fruitfulness, etc.) and which are acquired via training with paradigms. He writes,

> To many readers of the preceding chapters, this characteristic of the operation of shared values has seemed a major weakness of my position. Because I insist that what scientists share is not sufficient to command uniform assent about such matters as the choice between competing theories or the distinction between an ordinary anomaly, and a crisis-provoking one, I am occasionally accused of glorifying subjectivity and even irrationality. But that reaction ignores two characteristics displayed by value judgments in any field. First, shared values can be important determinants of group behavior even though the members of the group do not all apply them in the same way. . . . Second, individual variability in the application of shared values may serve functions essential to science. (Kuhn 1970, 185–186)

Kuhn's second point in this passage is an important reflection on group rationality and is connected to the evolutionary account of change he briefly introduces in the closing pages of SSR. Some processes, such as evolution, depend on variation, especially at times of stress. Hence it is helpful to progress for the standards applied not to fix a unique outcome but to permit a variety of possible views. Thus science must permit it to be rational for scientists, in some cases at least, to disagree without either being irrational. We must see individual rationality as bound up with the rationality of the group enterprise. And since the intuitions and similarity relations employed by scientists are shared, not merely individual, Kuhn can reject the charge of subjectivism.

The second feature of Kuhnian rationality is a matter of individual psychology: the central role of acquired intuitions about similarity, the ability of a scientist to recognize the similarity between a scientific puzzle and a paradigm, and likewise the similarity between their solutions. Kuhn does not attempt to articulate a detailed account of rationality on this basis, but the following, is, I suggest, how such an account might begin. A naturalized conception of rationality locates it not in dispositions to follow logical or other a priori rules but in capacities that fill a cognitive role. At a first approximation, a subject is acting rationally when she follows a cognitive procedure that meets the subject's cognitive goals. On this view, identifying people by recognizing their faces is an exercise of rationality (albeit a minimal one). Scientific rationality is a matter of acting on a similarity relation that contributes to the pursuit of science's cognitive goals, the solving of scientific puzzles. (So not every acquired similarity relation contributes to being rational.) As such, the rationality of a scientist is much more akin to the ability of an art connoisseur to recognize the work of an old master by its brushwork than it is to the ability of a logician to verify a proof using the rules of logic.[6] Similarity is not quantifiable but is learned by exposure to exemplars. Furthermore, the relevant features that go to make up a similarity relation are multiple and imperfectly commensurable. So there is no reason to suppose that (at most) a single conclusion is rationally mandated or that scientists in possession of the same evidence cannot rationally disagree (Kuhn 1970, 199).

We should note that because scientific rationality consists in exercising capacities that promote science's cognitive goals (maximizing the puzzle-solving power of science) there is scope for a clear demarcation between rational and irrational scientific behavior within Kuhn's picture. An irrational scientific choice occurs when one, other things being equal, prefers a hypothesis that fails to solve puzzles over one that does. And such choices will typically occur when the choices are influenced by factors external to science.[7] Thus Kuhn is able to make the same judgment of, for example, Lysenkoism that the logical empiricists would have, namely that it is degenerate, irrational science.

To summarize this section: Kuhn's theory requires him to take a predominantly internalist approach to explaining scientific change. This is true of revolutionary change as well as developments in normal science, Kuhn's few remarks about Kepler and sun worship and the like notwithstanding. This fact has been ignored by both Kuhn's critics and by proponents of externalist approaches in science studies. This is in part because both groups adopt a restricted, logical empiricist conception of rationality. Kuhn's account does indeed conflict with that conception, not because he regards scientists as irrational but because he is offering a different, more naturalistic conception of what rationality in science is.

3. THE STRONG PROGRAMME AND NATURALISM

Now I want to turn to the characteristic that Kuhn and at least one important strand in science studies do share. Kuhn's approach and that avowed by the Strong Programme are both naturalistic. By a "naturalistic" approach to some subject matter, I mean an approach that is willing to employ the methods and results of the natural sciences. If our subject matter is the nature of scientific thought and development, the approach of the logical empiricists was decidedly not naturalistic. It was almost entirely a priori. According to that view, the principal task for a philosopher of science is to find an a priori account of scientific rationality, be it an account of confirmation or of the logic of induction or, at least, of refutation. This, plus the minimal assumption of the rationality of most scientists, provides an account of what scientific thought in fact is. Likewise we have a schema for explaining, in the general case, why a scientist believes what he does believe—we refer to the evidence that the scientist had available, and the account of scientific rationality does the rest. Only in the peculiar cases, where a scientist is behaving irrationally, do we need to appeal to any additional factor and a form of explanation drawn from outside philosophy (e.g., a psychological or social explanation).

What was revolutionary about Kuhn was his willingness to apply social and most particularly psychological explanations in the general case (which gave further impetus to Kuhn's being perceived to be an irrationalist about science). We have discussed Kuhn's innovative naturalism above: science is driven by judgments of similarity to exemplary scientific puzzle solutions—paradigms-as-exemplars; an ability to make such judgments may be learned by practice with the paradigmatic exemplars and related kinds of example (such as textbook and exam questions). Such acquired dispositions have significant effects; they can affect one's perceptual judgments and more generally they channel one's thinking so that only puzzle solutions akin to the exemplars are likely to be spotted.

The sociology of scientific knowledge and the Strong Programme in particular are also avowedly naturalistic. SSK aims to use the methods and findings of science—social science—to understand the creation of scientific consensus (for which the word "knowledge" has become a term of art). David Bloor tells us,

> The search for laws and theories in the sociology of science is absolutely identical in its procedure with that of any other science. This means that the following steps are to be found. Empirical investigation will locate typical and recurrent events. Such investigation might itself have been prompted by some prior theory, the violation of a tacit expectation or practical needs. A theory must then be invented to explain the empirical regularity. This will formulate a general principle or invoke a model to account for the facts. (Bloor 1991, 21)

So the sociology of science employs the same general method as other sciences, the same naturalistic approach that I attributed to Kuhn in SSR in Section 2.

One of the problems, however, for SSK is how to distinguish itself from history of science. After all, the mere reference to social factors in explaining some change in scientific belief doesn't make the explanation sociological. Most historical explanations of ordinary historical events appeal to social factors. Barry Barnes (1990), in an article on sociological theories of scientific knowledge, mentions four studies that exemplify SSK by showing how specific objectives and interests within scientific communities play a role in explaining how consensus was formed on particular beliefs—Dean's work on theories of plant species, Pickering's on elementary particle models, MacKenzie's on statistical measurement, and Shapin's on the pictorial representation of the human brain. But the most avowedly externalist studies, those of MacKenzie and Shapin, do not strike one as manifestly distinct from historical studies that seek social explanations of scientific belief change, and Dean's work, while focusing to a greater extent on microsocial relations, is nevertheless largely historical in approach. Returning to the quotation from Bloor, these studies do not display laws and theories nor do they conspicuously formulate general principles or invoke models. Perhaps all that distinguishes them is that they see themselves as contributing to

SSK in the sense of giving examples of how social forces play a role in science, but that is not enough to make them distinctively sociological. What would make a study distinctively sociological would be if it employed some genuinely sociological theory, which is to say some contentful and general set of hypotheses about exactly how social forces play a role in science. If SSK could do that it really would be naturalistic, in the sense of employing the methods and results of science, rather than in the weak sense of just being empirical, which provides no distinction from history. Barnes seems to recognize this, and so cites Mary Douglas' grid-group analysis as an example of the sort of general theory I have mentioned, which would link kinds of social structure to kinds of belief. Yet such theories are few, and well-confirmed instances of such theories are even fewer.

Pickering's work is closer in nature to that of Bruno Latour, and like Dean's is concerned with microsocial explanations. Latour (1987) in particular might seem to exemplify Bloor's desire for the sociology of science. For he employs the naturalistic methods of anthropology and deploys a general explanatory theory, Actor-Network theory. Latour's microsociological work at first sight also appears less externalist than that of Shapin or MacKenzie. However, in the light of Latour's metaphysics, this is an illusion. "Social construction" for Barnes, MacKenzie, et al. refers to the construction of ideas and beliefs by social forces. But for Latour and also for Pickering, the very objects of scientific study are themselves social. In such a context it is difficult to make much sense of the idea of naturalism or of an internal-external distinction.

As we discussed above, before the 1960s a consensus existed that the explanation of the contents of belief must normally be internalist. This allowed for a demarcation between sociology and history of science thus: historians give internalist explanations of the content of science while sociologists give explanations, which would often be externalist, of the form (structure) of science. But with the breakdown of that consensus against externalist explanations of content, the demarcation between history of science and sociology of science becomes blurred. External explanation was hitherto exclusively the domain of the sociologists, so it was tempting to retain this as a characteristic feature of sociology, and to extend the reach of sociology to the content of science in addition to its form (we see this in the quotations from MacKenzie above). Yet the inability of studies of this kind, such as MacKenzie's, to distinguish themselves from externalist history, while not detracting from them as works of scholarship, does cast doubt on the success of SSK in developing a distinctively *sociological* study of science. Ironically, while few products of SSK come especially close to fulfilling Bloor's naturalistic aims as outlined in the quotation above, the work that best fits those aims is Kuhn's SSR. For it is naturalistic, and does identify patterns which it explains using a general theory/model that appeals to social forces. Yet, as we have discussed, those social forces are resolutely internal.

Is there then a different approach to the study of science, such that SSK could be a genuinely naturalistic and distinctive study of science? I believe that there is. Such an approach would, like Merton's project, concern itself with the structure and function of modern science. It would permit sociological explanations of why such a structure exists and what social forces help or hinder it. Like Kuhn's theory it would aim to be a general model of the structure of science, which when applied to particular episodes where scientific beliefs have changed would appeal in the normal case exclusively to the internal features of science. I suggest that this approach should take its cue from the work of Émile Durkheim (1893) and the functionalist school his thought gave rise to.[8] Thus we should ask what the *function* of science is, the answers to which question should explain the structure of science. And similar questions may be asked of the substructures of science.

Now some SSK practitioners may claim that they have answered the question about the function of science. Its function is to serve the objectives and interests of those engaged in science. Such a view is more or less explicit in Barnes' account of SSK. But I find such an answer unilluminating. It is the sort of explanation that can be applied to any institution. Of course, SSK may assert that is what science is, just another institution. But not all institutions are the same; they have different values (as Merton emphasizes), they have different internal structures, and they relate to other institutions and external structures in different ways. I believe that Durkheim's broader conception of function in sociology will help us answer such detailed questions concerning the nature of science. His analogy between society and an organism requires us to see the function of an institution as a matter of its interaction with other institutions and parts of society so that together they contribute to the proper functioning of the whole. This analogy has proved fruitful in sociology in general and I believe that it may be particularly fruitful in relation to the sociology of science.

Many biological organisms have organs or systems whose function is cognitive—which is to say that they are supposed to give the organism information about its environment, thereby influencing its behavior. One might suppose that social "organisms" might similarly possess institutions that perform a cognitive function. Certainly that is how one should understand military intelligence and signals corps in relation to the armed forces, and one may consider that publicly organized science plays a cognitive function with respect to society at large. The biological analogy is at best only suggestive here—one should not suppose that every biological organ (the liver, the pancreas, etc.) has a sociological analogue. But the fact that science does exist as an institution with an avowedly cognitive role adds weight to the suggestion, as does the fact, as social epistemologists have recognized, that there is an irreducibly collective notion of knowledge, which is used in particular in connection with science, and which is best understood as a social analogue of individual knowing (Corlett 1996; Bird 2010).

Such an approach has been ignored not only because sociological explanations have hitherto been interest-focused but also because the sociology of science has eschewed use of notions such as "truth" or "knowledge" (in the sense of "knows" that implies that what is known is true). This is in part motivated by an entirely reasonable, methodological preference, expressed in the symmetry principle, that the truth or falsity of some specific scientific belief should not appear in the explanation of why it is held. This is reasonable since the most immediate and informative explanation of a scientist's belief involves the evidence she possesses, not the fact that may be the ultimate cause of that evidence. However, the disdain for the concept of truth in SSK typically goes somewhat further than this, partly because of skepticism about the accessibility of truth and partly because of more general relativist pressures.[9] But if one refuses to avail oneself of standard notions of truth and knowledge, it becomes impossible to think of an institution as having a cognitive function. The very point of a cognitive function is to deliver reliable information, to provide a link with the truth, and to filter out what is false. On the other hand, if one can talk about the truth or knowledge-seeking function of science then one is in a position to examine the structure of science with a view to identifying its truth-tropic character and more interestingly, the pathologies from which it may suffer. (The fact that some thing may be a pathology, such as fraud or the distortion caused by commercial science, does not mean that it is rare.) For example, the world's leading medical journals have agreed to publish only the results of studies that have been registered prior to their being carried out. The aim of this is to combat the problem of selective publication of results. One could see this simply as some sort of power struggle between journal editors and pharmaceutical companies. But it is much more enlightening to see this as a corrective to a pathology. It strikes me that sociologists could usefully take more interest in such developments.

Of course, if the Strong Programme is genuinely naturalistic then it ought not issue a blanket prohibition against employing the concept of truth or against any view that treats in certain cases true beliefs and false beliefs differently. David Bloor explicitly sets up the Strong Programme against those who assume a priori that true beliefs and false beliefs require different forms of explanation. But why is it any better to assume a priori that they must always be treated the same? Of course this opposition is more subtle than Bloor sets it up to be, since no one, not even Lakatos, supposed that *every* false scientific belief requires a special kind of explanation different from true ones. No traditionalist supposed that Newton's belief in action at a distance or in his own equations of motion, or Maxwell's belief in the aether, or Dalton's belief that water is HO, and so forth, requires some special kind of explanation reserved for science that has gone awry, even though these are all false beliefs. So the distinctions among different kinds of belief and different explanations of them was always more subtle than the distinction between the true and the false. A naturalist ought to regard it as an empirical question whether there are any interesting distinctions among kinds of belief and their explanations, and it ought to be a matter of discovery whether truth or rationality play no part, or some part, or a major part in those explanations. As a corrective to the weak programme a bald symmetry principle may be useful. But if we are going to be more careful, a naturalistic symmetry principle ought to be stated thus: do not adopt a priori (i.e., *unless supported by appropriate evidence*) any principle or method that assumes that true and false beliefs must be treated differently. Whether or not one adopts this particular more liberalized version of the symmetry principle, the Strong Programme should, I suggest, be seen as a metamethodological standpoint, telling us how we should choose our methods and explanatory principles. Understood thus its methodological relativism can be seen to align with naturalism: allow empirical, sociological investigation to reveal whether we (the analysts) need to make use of categories such as "truth" and "knowledge." This methodological relativism rejects philosophical relativism which gives a nonempirical, a priori (negative) answer to this question.

Finn Collin (2011) argues that science and technology studies (STS) has failed to live up to its avowed empirical, naturalistic aims, in his view because of the very intensity of its agenda against science and traditional philosophy of science. According to Collin the need to show that science's susceptibility to social forces is *inevitable* has led Bloor and Barnes to deploy *philosophical* arguments to support their conclusions and so retreat from their stated empiricist commitments. I have mentioned Barnes' use of the Duhem–Quine underdetermination argument. Prominent in the work of both Barnes and Bloor and also Kusch and others is a finitist theory of meaning that draws upon Wittgenstein's private language argument. It is certainly striking that philosophical theory is more prominent than sociological theory in Bloor, Barnes, and Henry's *Scientific Knowledge: A Sociological Analysis* (1996). Collin himself advocates a new role for STS, what he calls *critical sociology of knowledge society*, by analogy with classical sociology conceived of as the science of industrial society. The sociology of knowledge society would ask important and critical questions about the nature of society in which knowledge plays such a central role, for example: Is knowledge society still a class society and, if so, which are its classes? Does a liberal, democratic society offer the best conditions for the generation of scientific knowledge and for its transformation into new consumer goods, produced according to innovative, knowledge-intensive methods? Is the classical Leninist theory of colonialism still valid, or has some parity between the old industrialized world and the "third world" been achieved through the outsourcing of highly skilled job functions in the new "knowledge economy"? While Collin's proposal has a different focus from mine, it strikes me that these questions are at least consistent and may form a coherent whole. For Collin says that a reformed STS would try to identify those social mechanisms that make science more responsive to reality. Like my proposal, it would be "veritistic" in Goldman's sense (1999), taking *truth* seriously as an explanatory concept (Collin proposes realism as a methodological stance, whose fruitfulness would be evaluated empirically). It is worth noting that some of what has been suggested will not be entirely against the grain of recent STS. Collins and Evans (2002) call for a "third wave" in science studies which recognizes "expertise" as a real, analyst's category, which is to say that "expert" is not just an actor's category (used to designate persons taken *by the groups being studied* to fulfil a certain function) but is a category used by the sociologist to categorize persons with particular objective attributes. This is important, for while it does not directly rehabilitate "truth" and "knowledge" as analyst's categories, the implication is that categories of this sort have an objective reality that the sociologist needs to exploit.

4. THOMAS KUHN'S NATURALISTIC LEGACY

I started this article by noting two (related) strands in Kuhn's naturalism: a historical-sociological one and a cognitive-psychological one. While SSK has sought to carry forward the former, I have suggested that it has failed to do so because of its overemphasis on externalist explanations of scientific belief and a failure to commit fully to naturalism, in particular by equivocating between methodological and a priori relativism. By contrast I want to suggest that there is a growing body of research that draws rather more effectively upon the second strand of Kuhn's naturalism. In this section I shall give a very brief overview of this new direction for Kuhn's ideas.

For example, in a paper entitled "How Do Scientists Think? Capturing the Dynamics of Conceptual Change in Science" Nancy Nersessian (1992) introduces "cognitive-historical analysis," which starts from the assumption that the representational processes and problem-solving strategies employed by scientists are highly sophisticated and refined versions of processes and strategies in more general use. It therefore combines historical case studies of actual episodes with insights from psychology and cognitive science to better understand scientific thinking. Nersessian acknowledges a Kuhnian backdrop to this work, which has the additional merit of explicitly linking the psychological and historical strands I have been discussing.

The programs of research that connect to psychological aspects of the Kuhnian notion of paradigm themselves tend to fall into two (connected) categories. The first emphasizes the *cognitive* aspects of paradigms and the second emphasizes the *conceptual*. Roughly, the distinction corresponds to those who wish to further a naturalistic explanation of scientific development (problem solving, theory change, and the like) and those who are focused more on a naturalistic understanding of incommensurability and conceptual change.

Nersessian's work (1987, 2003), while embracing both directions, has tended to concentrate more on the second, conceptual side. Andersen,

Barker, and Chen (1996, 1998, 2006) in particular have progressed this line of thought. Kuhn argues that exemplars fulfill a conceptual function, the extension of a concept being determined by similarity to a set of exemplary cases rather than by an intension. Andersen, Barker, and Chen propose that this approach is supported by empirical work (Rosch 1973; Rosch and Mervis 1975) on the use of prototypes in category formation. They argue that this can be used to give a naturalistic explanation of semantic incommensurability, via an understanding of dynamic frames.

In parallel, philosophers have drawn on the work of psychologists in further articulating Kuhnian ideas concerning the nature of paradigms and paradigm change. Howard Margolis' (1987) *Patterns, Thinking, and Cognition* was an early work in this direction. Tom Nickles (2003) has noted the link between Kuhn and recent work in case-based reasoning (cf. Leake 1998), while I (2005) have related this to psychological work by Kevin Dunbar (1996, 1999) on analogical thinking by scientists.

The naturalistic Strong Programme should therefore be in good company. If we are looking for a Kuhnian legacy, this naturalistic pursuit of Kuhnian themes is very much in the manner of SSR. But, as I have suggested, it is not clear that SSK has in general taken full advantage of the opportunities that naturalism offers. A telling contrast is between the work of Dunbar and that of Bruno Latour. Both use ostensibly the same methods—microinvestigation of the activities of scientists within a laboratory. Dunbar spent several years working in leading microbiological laboratories in the United States, Canada, and Italy, filming their work, lab meetings especially, and recording interviews. Dunbar was interested in sociological issues, such as how the organization of a laboratory influences its effectiveness. But his principal interest was the ways in which scientists actually reason. His analysis revealed several types of inference all of which are analogical to a greater or lesser degree, at once giving us interesting general conclusions about the nature of scientific thought and also providing empirical confirmation of Kuhn's account of scientific thought in terms of exemplars. In contrast Latour and Woolgar (1986) show an interest in how scientists reach their scientific conclusions only insofar as this exhibits their techniques of persuasion and negotiation, rather than as mechanisms of cognition. In this respect, Dunbar is an heir to Kuhn's SSR whereas Latour is not.

The naturalism in Kuhn thus has an historical-sociological element and a cognitive-psychological element, and in science studies the former has been given precedence to the almost total exclusion of the latter. Both are necessary for Kuhn's model; within that model the psychological element is causally primary and the social element is secondary. The role of social forces in scientific training in particular but also in career success, changing paradigms, and so forth depends on the possibility of the psychological function of exemplars. A scientific education inculcates what Ludwik Fleck (1979) called a "thought-style" by training with a recognized set of exemplary puzzle solutions. In normal science the conservative consensus is a consensus on exemplars. In revolutionary science what matters is the ability of scientists to alter their thought styles. The psychology of a thought style means that it resists change. But it is not absolutely resistant to change, especially in younger scientists among whom the thought style has not taken such deep root. Given these kinds of explanation it is difficult to see how one could satisfactorily understand the historico-social significance of exemplars without understanding their psychological function.

5. CONCLUSION

All in all, the link between Kuhn and much of contemporary science studies has been greatly exaggerated. Externalist explanations, which are simply inconsistent with a Kuhnian model of scientific development, predominate in the latter. As regards the Strong Programme in particular there is a common naturalistic approach, but even here there is a divergence. Kuhn's theory and his naturalism should lead one to focus on psychology as much as on sociology.

Furthermore, I suspect that the outlook for the sociology of scientific knowledge is limited so long as it retains too strong a commitment to the externalism Kuhn rejected and to a version of the symmetry thesis that requires us to ignore truth altogether. So long as that remains we will be unable to ask important questions concerning the function and structure of science that once were and ought again to be central to the sociology of science.[10]

NOTES

1. Barnes (1981, 493; 1990, 63–64) is a particular example. He also attempts to pin this on Kuhn, referring to "Kuhn's famous dramatisation of the insufficiency of reason and experience in scientific evaluation" (1990, 64).
2. I believe that Laudan (1990) hugely overemphasizes Kuhn's commitment to underdetermination. In any case, to the extent that there is *some* underdetermination during revolutionary science, that is a consequence of the account, not an assumption, whereas for Barnes, underdetermination is an assumption taken from philosophers of science.
3. Kuhn (1992, 9) makes a related but slightly stronger point, which he attributes to Marcello Pera, that "the authors of microsociological studies are . . . taking the traditional view of scientific knowledge too much for granted."
4. This is how Roy Porter (1990, 37), for example, sees things. As regards the social determination of the content of science, he writes, "Of crucial importance in this respect—not necessarily for what it overtly stated but for what were taken to be its implications—was Thomas S. Kuhn's *The Structure of Scientific Revolutions*."
5. This is an important point—Kuhn frequently appeals to what it is rational for scientists to believe or do, but nowhere says that science is not a rational enterprise—but a point that is ignored by sociologists of science who appeal to Kuhn, for example. Barnes does acknowledge that it is a misconstrual of

Kuhn to see him as attacking science; but it is clear that to see him as reject-ing the rationality of science is also a misconstrual.

6. Two further points may be made here. The ability of the mathematician to *construct* the proof may depend on similarity relations acquired by training and experience. And even the logician's ability to see that a step in the proof instantiates a rule might be seen as another example of pattern recognition. So perhaps we should not contrast rule following and acquired similarity relations but instead regard rule following as just a special case of applying similarity relations. See Bird (2005).

7. Often things will not be equal in extraordinary science, for example because of Kuhn loss. Consequently, it will be possible, as emphasized above, ratio-nally to continue to resist a revolution, as Kuhn remarks is true of Priestley. Nonetheless, the room for being a rational holdout will diminish over time as the promise of a new paradigm is converted into proven success.

8. Barnes et al. (1996) also use Durkheim's ideas—but a different subset thereof. They refer to Durkheim's *The Elementary Forms of the Religious Life* (1912) in order to draw an analogy between science and religion. Durkheim's key distinction between the sacred and the profane, they say, explains why there is resistance to the naturalistic study of science itself. But I think this fails to appreciate the function of religion, as Durkheim understands it, which is to be a source of social solidarity, causing the individual to put the group's interests ahead of his own. Science does not have this function, and so there is no reason, in Durkheimian terms, why the sacred-profane distinction should be applied to science.

9. In SSR Kuhn talks little about truth, as if he is obeying a methodological symmetry principal. However in the Postscript to the second edition, Kuhn (1970, 206–207) does succumb to a skeptical argument of a Kantian kind; see Bird (2000, 225–232).

10. I am grateful for the very helpful comments of Vasso Kindi and Theodore Arabatzis on a draft of this paper. Research for this project was supported by a Research Fellowship funded by the Arts and Humanities Research Council (Grant AH/I004432/1).

REFERENCES

Andersen, H., P. Barker, and X. Chen 1996. "Kuhn's Mature Philosophy of Science and Cognitive Psychology." *Philosophical Psychology* 9: 347–363.

Andersen, H., P. Barker, and X. Chen 2006. *The Cognitive Structure of Scientific Revolutions*. Cambridge: Cambridge University Press.

Barnes, B. 1981. "On the 'Hows' and 'Whys' of Cultural Change (Response to Woolgar)." *Social Studies of Science* 11: 481–498.

Barnes, B. 1990. "Sociological Theories of Scientific Knowledge." In R. C. Olby, G. N. Cantor, J. R. R. Christie, and M. J. S. Hodge (eds.), *Companion to the History of Modern Science*. London: Routledge, 60–73.

Barnes, B., and D. Bloor 1982. "Relativism, Rationalism and the Sociology of Knowledge." In M. Hollis and S. Lukes (eds.), *Rationality and Relativism*. Blackwell, 21–47.

Barnes, B., D. Bloor, and J. Henry 1996. *Scientific Knowledge: A Sociological Analysis*. London: Athlone Press.

Barsalou, L. W. 1992. "Frames, Concepts, and Conceptual Fields." In A. Lehrer and E. F. Kittay (eds.), *Frames, Fields, and Contrasts: New Essays in Semantic and Lexical Organization*. Hillsdale, NJ: Lawrence Erlbaum Associates, 21–74.

Ben-David, J. 1971. *The Scientist's Role in Society: A Comparative Study.* Englewood Cliffs, NJ: Prentice Hall.

Bird, A. 2000. *Thomas Kuhn.* Chesham: Acumen.

Bird, A. 2005. "Naturalizing Kuhn." *Proceedings of the Aristotelian Society* 105: 109–127.

Bird, A. 2010. "Social Knowing." *Philosophical Perspectives,* 24: 23–56.

Bloor, D. 1991. *Knowledge and Social Imagery.* 2nd ed. Chicago: University of Chicago Press.

Chen, X., H. Andersen, and P. Barker 1998. "Kuhn's Theory of Scientific Revolutions and Cognitive Psychology." *Philosophical Psychology* 11: 5–28.

Collin, F. 2011. *Science Studies as Naturalized Philosophy.* Dordrecht: Springer.

Collins, H. 2004. *Gravity's Shadow: The Search for Gravitational Waves.* Chicago: University of Chicago Press.

Collins, H. M., and R. Evans 2002. "The Third Wave of Science Studies." *Social Studies of Science* 32: 235–96.

Corlett, J. A. 1996. *Analyzing Social Knowledge.* Lanham, MD: Rowman and Littlefield.

Dunbar, K. 1996. "How Scientists Really Reason." In R. Sternberg and J. Davidson (eds.), *The Nature of Insight.* Cambridge, MA: MIT Press, 365–395.

Dunbar, K. 1999. "How Scientists Build Models: In Vivo Science as a Window on the Scientific Mind." In L. Magnani, N. J. Nersessian, and P. Thagard (eds.), *Model-Based Reasoning in Scientific Discovery.* New York: Kluwer/Plenum, 85–99.

Durkheim, É. 1893. *De la division du travail social.* Paris: Alcan. Translated as *The Division of Labor in Society,* by W. D. Halls, New York: The Free Press, 1984.

Durkheim, É. 1912. *Les formes élémentaires de la vie religieuse.* Translated as *The Elementary Forms of the Religious Life,* by Joseph Ward Swain, New York and London: The Free Press, 1915.

Farley, J., and G. Geison 1974. "Science, Politics and Spontaneous Generation in Nineteenth-Century France: The Pasteur–Pouchet Debate." *Bulletin of the History of Medicine* 48: 161–98.

Fleck, L. 1935/1979. *Genesis and Development of a Scientific Fact.* Chicago: University of Chicago Press.

Fuller, S. 2002. "With Friends like This, Who Needs Enemies?" *Metascience* 11: 46–51.

Goldman, A. I. 1999. *Knowledge in a Social World.* Oxford: Clarendon Press.

Hessen, B. M. 1931/1971. "The Socio-Economic Roots of Newton's Principia." In N. Bukharin (ed.), *Science at the Crossroads.* Papers presented to the International Congress of the History of Science and Technology, 1931, by the Delegates of the U.S.S.R. London: Frank Cass.

Kuhn, T. S. 1962. *The Structure of Scientific Revolutions.* Chicago: University of Chicago Press.

Kuhn, T. S. 1968. "The History of Science." In *International Encyclopedia of the Social Sciences,* vol. 14. New York: Crowell Collier and Macmillan, 74–83. Page references to T. S. Kuhn, *The Essential Tension.* Chicago: University of Chicago Press, 1977.

Kuhn, T. S. 1970. *The Structure of Scientific Revolutions.* 2nd ed. Chicago: University of Chicago Press.

Kuhn, T. S. 1971. "The Relations between History and the History of Science." *Daedalus* 100: 271–304. Page references to T. S. Kuhn, *The Essential Tension.* Chicago: University of Chicago Press, 1977.

Kuhn, T. S. 1977. "Objectivity, Value Judgment, and Theory Choice." In T. S. Kuhn, *The Essential Tension.* Chicago: University of Chicago Press, 320–339.

Kuhn, T. S. 1978. *Black-Body Theory and the Quantum Discontinuity, 1894–1912*. Chicago: University of Chicago Press.

Kuhn, T. S. 1992. "The Trouble with the Historical Philosophy of Science." Robert and Maurine Rothschild Distinguished Lecture, 19 November 1991. An Occasional Publication of the Department of the History of Science. Harvard University Press, Cambridge, MA.

Lakatos, I. 1970. "Falsification and the Methodology of Scientific Research Programmes." In I. Lakatos and A. Musgrave (eds.), *Criticism and the Growth of Knowledge*. Cambridge: Cambridge University Press, 91–195.

Latour, B. 1987. *Science in Action: How to Follow Scientists and Engineers through Society*. Cambridge, MA: Harvard University Press.

Latour, B., and S. Woolgar 1986. *Laboratory Life: The Construction of Scientific Facts*. Paperback ed. Princeton, NJ: Princeton University Press.

Laudan, L. 1990. "Demystifying Underdetermination." In C. W. Savage (ed.), *Scientific Theories*, vol. 14 of Minnesota Studies in the Philosophy of Science. Minneapolis, MN: University of Minnesota Press, 267–297.

Leake, D. 1998. "Case-Based Reasoning." In W. Bechtel and G. Graham (eds.), *A Companion to Cognitive Science*. Oxford: Blackwell, 465–476.

MacKenzie, D. A. 1981. *Statistics in Britain, 1865–1930: The Social Construction of Scientific Knowledge*. Edinburgh: Edinburgh University Press.

Margolis, H. 1987. *Patterns, Thinking, and Cognition. A Theory of Judgment*. Chicago: University of Chicago Press.

Merton, R. K. 1938. "Science, Technology and Society in Seventeenth Century England." *Osiris* 4: 360–632.

Nersessian, N. 1987. "A Cognitive-Historical Approach to Meaning in Scientific Theories." In N. Nersessian (ed.), *The Process of Science*. Dordrecht: Kluwer, 161–177.

Nersessian, N. 2003. "Kuhn, Conceptual Change, and Cognitive Science." In T. Nickles (ed.), *Thomas Kuhn*. Cambridge: Cambridge University Press, 179–211.

Nersessian, N. J. 1992. "How Do Scientists Think? Capturing the Dynamics of Conceptual Change in Science." *Cognitive Models of Science* 15: 3–44.

Nickles, T. 2003. "Normal Science: From Logic to Case-Based and Model-Based Reasoning." In T. Nickles (ed.), *Thomas Kuhn*. Cambridge: Cambridge University Press, 142–177.

Nickles, T. 2012. "Some Puzzles about Kuhn's Exemplars." In Kindi and Arabatzis (eds.) *Kuhn's The Structure of Scientific Revolutions Revisited*. New York: Routledge, 112–133.

Porter, R. 1990. "The History of Science and the History of Society." In R. C. Olby, G. N. Cantor, J. R. R. Christie, and M. J. S. Hodge (eds.), *Companion to the History of Modern Science*. London: Routledge, 32–46.

Quine, W. V. 1969. "Epistemology Naturalized." In *Ontological Relativity and Other Essays*. New York: Columbia University Press, 69–90.

Rosch, E. 1973. "On the Internal Structure of Perceptual and Semantic Categories." In T. E. Moore (ed.), *Cognitive Development and the Acquisition of Language*. New York: Academic, 111–144.

Rosch, E., and C. B. Mervis. 1975. "Family Resemblances: Studies in the Internal Structure of Categories." *Cognitive Psychology* 7: 573–605.

Sulloway, F. 1996. *Born to Rebel: Birth Order, Family Dynamics, and Revolutionary Genius*. New York: Pantheon.

Toulmin, S. 1970. "Does the Distinction between Normal and Revolutionary Science Hold Water?" In I. Lakatos and A. Musgrave (eds.), *Criticism and the Growth of Knowledge*. Cambridge: Cambridge University Press, 39–47.

11 The Structure of Philosophical History
Thoughts after Kuhn[1]

Alan Richardson

Thomas Kuhn's *Structure of Scientific Revolutions* (hereafter, SSR) is one of the most important academic books of the last century. Viewing it at the distance of a half century, one can say with confidence that Kuhn fundamentally altered perceptions about science and its history, not merely among professional historians but also among practicing scientists, philosophers of science, a variety of scholars in many disciplines, and a large lay audience. In ways Kuhn himself did not always understand or appreciate, his work gave new or renewed life to fields such as sociology of science, rhetoric of science, and cultural studies of science. Philosophy of science was also importantly altered—Kuhn is widely understood to have been the largest figure in the movement away from logical empiricism and toward a new and richer, if more difficult to characterize, phase of philosophical understanding of science. Philosophers took up Kuhn early and with characteristic vehemence. Some philosophers of science such as Karl Popper early on found Kuhn's account of science not merely wrongheaded but dangerous. Other philosophers, such as Richard Rorty, by the end of the 1970s were keen to co-opt Kuhn's work for their own purposes, however far those purposes might have been from Kuhn's. The philosophical community played a large role in shaping Kuhn's development after SSR, as he was drawn into discussions of meaning and reference, realism, and rationality.[2]

This essay discusses none of these topics directly. I am not interested here in the specifics of Kuhn's account of the historical development of science. I am, rather, interested in how, in SSR, Kuhn motivated a new interest in the history of science and whether one could, as an homage fifty years later, spark a new interest in the history of another field on essentially these same Kuhnian grounds. I mean to discuss the relations of philosophy to its history and to use the first chapter of SSR to invite curiosity regarding the entanglement of tasks assigned to the historian of philosophy and overall narratives of the historical development of philosophy. Kuhn serves me here as a guide and an inspiration, but I do not merely acknowledge but insist on the fact that Kuhn himself offered no great insight into the history of philosophy (nor did he attempt to). My efforts here, however slight, are offered in the spirit of Kuhn's inquiry, not as an argument for or against any specific claim he made within that inquiry. I hope to show that fifty years on we can find in his work new and perhaps surprisingly rich insight into fields relatively far from Kuhn's own interests. The essay is an invitation to think anew about the relations of philosophy and its history; while it relies on some historical work I and others have done, it more calls for than provides the sort of rigorous historical scholarship that Kuhn both demanded and provided.

A ROLE FOR HISTORY: CHAPTER 1 OF SSR

Kuhn's SSR is not in the first instance an essay in philosophy of science. It is an essay in the historiography of science. Kuhn begins the book by describing the questions typically taken up by the historian of science. He then argues that these questions are, when answerable, boring and, when interesting, unanswerable. It is in his diagnosis of why the questions are unanswerable in the interesting cases that leads Kuhn to outline the "image of science by which we are possessed" (SSR, 1). This image is (in 1962), according to Kuhn, both widely available in science textbooks and largely codified by philosophers of science; this last claim is why ultimately Kuhn's attempts to breathe life into the work of the historian of science had to engage with philosophy of science.[3]

According to Kuhn, the questions bequeathed to the historian of science by the standard image of science and of its progress through history have two typical features. They presuppose a cumulative progress of science—and thus invite the historian to investigate where and when a certain bit of our current stock of scientific beliefs, theories, experimental techniques, or what have you was first added to that stock. They also are robustly individualistic, inviting curiosity about which bits of optical experimentation or theory, for example, came from the work of individual scientists such as Snell, Descartes, Huygens, or Newton. This twofold structure of typical historical questions yields investigations seeking to answer questions like these: Did Lavoisier discover oxygen? What was Rosalind Franklin's role in the discovery of the structure of DNA? Additionally, the cumulative view of scientific progress presents the historian with an opportunity to discuss impediments to progress—and it is here that larger cultural factors can be cited. Thus, the historian can ask what social factors prevented modern combustion theory from arising in England or what caused Linus Pauling to fail in his effort to discover the structure of DNA. Here is how Kuhn puts the point (SSR, 2):

> Concerned with scientific development, the historian then appears to have two main tasks. On the one hand, he must determine by what man [*sic*] and at what point in time each contemporary fact, law, and theory was discovered or invented. On the other hand, he must describe and

explain the congeries of error, myth, and superstition that have inhibited the more rapid accumulation of the constituents of the modern science text.

Kuhn points out an oddity for questions of these sorts: the more one knows about the historical period under scrutiny and the more one understands the historically available theories at that time, the harder it is to answer such questions. Thus, the question about Lavoisier's role in the discovery of oxygen hides within it the whole question of exactly what discovering oxygen consists in. Is it a matter of someone somehow purposefully isolating a relatively pure sample of what we now call "oxygen" even if the scientist in question called it "dephlogisticated air"? Or does the person need also to have articulated something close enough to an oxygen theory of combustion for us to think that he really was discovering a chemical element, oxygen, rather than the absence of something else? Similarly, the more one understands the intellectual context and content of, say, Aristotle's theory of motion, the less one sees the issue separating his theory from ours (or Newton's) to be one of error, myth, or superstition that prevented Aristotle from theorizing properly and more a matter of deep conceptual divides between our concept of motion and his. If honest, rational, methodical people have, in their systematic inquiry into nature, managed to come up with theories deeply incompatible with ours, then the idea of progress through accumulation is rendered suspect—what separates us from Aristotle is conceptual change, not simply the accumulation of fact or elimination of irrationality and myth.

Such problems led Kuhn to claim that there was in his time the beginnings of an "historiographic revolution in the study of science" (SSR, 3), a revolution that suggested a rival image of science. Indeed, he claimed that SSR aimed "to delineate that image by making explicit some of the new historiography's implications" (3). This image is the now famous image of science as proceeding largely in periods of paradigm-driven normal science—a radical problem-solving tradition that brings the problem-solving capabilities of that paradigm to the limit—, the ensuing conceptual crisis, and then reorganization conceptually and socially in periods of revolution. This Kuhnian image of science renders at least highly problematic any account of science as a presuppositionless project of getting closer and closer to the truth about nature, standing in splendid isolation from other parts of culture, and finding its way through its conceptual maze without help from beyond its well-policed borders.[4] It also rendered implausible any effort to read the social organization of liberal democracy from the open society of scientific researchers engaged in searching criticisms of one another's work.[5]

As I say, I am not particularly interested here in the details of Kuhn's image of science. I am interested in how Kuhn sets up his problem in the first chapter of SSR. The argument structure of the first chapter of SSR has received much less comment than it merits, given how striking it really is: Historians have had difficulties with the questions assigned to them by an image of science given expression in science textbooks and codified in philosophy of science. When called upon to answer those questions, historians discover that they are either remarkably difficult to answer or seem to be historically unilluminating, even deeply misleading. Thus, the effort by historians to ask and answer better questions is the motivation for a new image of science. This new image is offered despite the commitment of scientists themselves to the received image. The very idea that scientists' conception of science might be changed by the work of historians of science is wonderful hubris, made all the more wonderful by the fact that it has been strikingly successful.

HISTORY OF PHILOSOPHY: SOME ROLES

Might it be possible to extend Kuhn's argument into a different realm? Could close attention to the difficulties attending upon attempts to answer the historical questions in another field of human endeavor provide a motivation to change our view of that field? Some fields suggest themselves. Technology, for example, seems very much to be immersed in a narrative of cumulative progress and individual genius. Perhaps it too is beset with the problems such a narrative poses for historical understanding.

Philosophy is another field that suggests itself. This is not because philosophy is immersed in a narrative similar to the one that Kuhn found had hobbled attempts to do history of science. Not only is philosophy not subject to an image of perpetual cumulative progress at the hands of individual geniuses, but also arguably there is no clear and consistent "image of philosophy" by which anyone, including the professional philosopher, is possessed. Instead, we have competing narratives of the overall structure of philosophical history, narratives neither clearly articulated nor fully believed by working philosophers or historians of philosophy. These narratives are most usually visible in introductory lectures to skeptical beginning students—philosophy as engaged in the endless human struggle with the largest issues of life or philosophy as the birth mother of other disciplines. Nonetheless, these narratives might very well constrain what is expected of historical scholarship regarding philosophy. There are other interesting contrasts to the case of science. For example, when Kuhn wrote SSR there was already an established professional community of historians of science distinct from the community of scientists. While there are historians who work on history of philosophy, most historians of philosophy are trained philosophers, working in philosophy departments. There is, however, role differentiation within philosophy, with many philosophers never doing historical research and many philosophers publishing only in history of philosophy. Another difference between science and philosophy—one perhaps not unrelated to the previously mentioned differences—is that history of philosophy is understood to be an important part of training within philosophy all the way up through the doctoral level; very rarely are scientists expected to take courses or undertake research in the history of science. Indeed, it is often noted that philosophy is unique among disciplines in requiring training in its history of just about every student who wishes to take a degree in the subject.

Notwithstanding this emphasis on history in the curriculum and the professional standing within philosophy of scholars working in history of philosophy, it has also been claimed that philosophy is among the least historically minded of all disciplines. Indeed, it is not uncommon for trained historians to puzzle at what counts as history of philosophy among philosophers. It is not uncommon for a philosophy graduate course in, say, the work of G. W. Leibniz, to read a very small portion of his writing—all Leibniz's work in mathematics, history, and so on having been carefully excised—, to read none of the work Leibniz was responding to, to make no reference to the events in Leibniz's life, to contemporaneous scientific discoveries (even those of Leibniz himself), nor to the larger intellectual or political culture in which he found himself, and to read as secondary literature only very recent commentaries on the very works of Leibniz. Such an approach to the text of Leibniz's writings can seem absurdly unhistorical—none of Leibniz's motivations and none of the historical context of his work are explored, no effort is made to understand the place of philosophy within his intellectual work as a whole. Moreover such courses presuppose rather than answer the question of why Leibniz is a particularly important person to study in the history of philosophy. Indeed, such courses often tip from the unhistorical into the antihistorical, when, for example, the excisions are made not on the understanding of what was philosophy at the time of the work being read but on what is now understood to be philosophy. For example, often the convergence of science and natural philosophy in the seventeenth century is waived and, thus, all of Leibniz's interventions into physics are ignored even though he did not think he was ceasing to be a philosopher in making those interventions. Almost certainly the bridges Leibniz built between his technical work in mathematics and the detailed metaphysics of monads will be at best gestured at—more often they are ignored, and perhaps most often they are not even noticed. In result, then, we end up with, as Bruce Kuklick (1984) has remarked, courses aimed at answering the question of what certain historically great philosophers had to say on what we now take to be philosophy—with even the question as to why these figures and not others are historically great taken as answered simply in the fact that these are the philosophers we still read. What situation could be more harmful to historical understanding of the development of philosophy?

Now, as already mentioned, what makes the situation more complicated than the one Kuhn found regarding science and history of science is the absence of anything that might be called "an image of philosophy by which we are possessed." There are, however, as also mentioned above, a couple of narrative structures that are brought in on occasion to frame the historical development of the discipline. One of these narratives might be placed under the slogan "Philosophy as fertile mother." According to this narrative, philosophy is the birthplace of other disciplines as they make their first steps toward being sciences in their own right. Another narrative is called "philosophia perennis." According to this narrative, philosophy is the discipline that constantly considers certain perennial questions. A favored list of questions involves the meaning and importance of key notions like truth, beauty, justice, the good, knowledge, and so on.

The fertile mother narrative is favored often by those who wish to find a place for philosophy in the progress of science while not insisting that philosophy itself has a scientific status. According to this narrative, in various places and times, philosophy has managed to investigate key concepts in ways that make them finally amenable to scientific investigation. The seventeenth century may have done this for various physical concepts— mass, force, motion, time—while psychological notions had to wait until the eighteenth or nineteenth century before they moved from notions of prima philosophia to empirical psychology. Philosophy on this view earns its keep by being the origin of scientific disciplines.

The historiographic difficulties here are obvious: Suppose philosophy considers a set of concepts—psychological concepts, for example—for over two thousand years. Why would these concepts become ripe for empirical scientific consideration in the eighteenth or nineteenth century? More importantly, would this development be properly understood as a development within philosophy—philosophy is finally able to give birth to this scientific discipline in Leipzig in 1880, say—or is it more likely that it is an external development—perhaps a self-conscious attempt to find empirical, not philosophical, methods for studying the mind, or the development of a new set of measuring instruments, or what have you? We need not answer these questions in any detail for any particular discipline; we need only notice the sorts of questions raised on this narrative structure of philosophy: why did this field move from philosophy to science at the place and time that it did? And: is this movement something that is properly called "birth" or properly called "repudiation"?

Of course, any origin story in which one discipline gives rise to another—chemistry from alchemy or astronomy from astrology or what have you—needs to answer specific questions of why and when this happened. What is unique in the narrative structure of philosophy as fertile mother is that it posits an integrated field of study that maintains both its identity and its fecundity across centuries of such birth events. Alchemy ceased to be a serious intellectual project precisely in and through the process by which it changed into chemistry. Astrology became a marginal science and ultimately a farce as astronomy clearly split from it. Philosophy, on the other hand, can give birth to mathematics in the ancient world, to physics in the seventeenth century, to psychology and economics in the nineteenth century, and yet still be available to teach as an ongoing intellectual pursuit in the twenty-first century. How is this possible?

I suspect that on this narrative structure, we do have a situation very close to the situation Kuhn found in history of science: the questions here are unanswerable, misposed, or uninteresting. Indeed, the entire narrative is based on a strict separation between philosophy and science—while the development of that very separation is among the most important questions, it seems to me, a history of Western philosophy ought to explain. So, did Newton cause physics to move from philosophy to science even though he himself thought of physics as "natural philosophy"? Do we have to dig around in the prehistory of every discipline, hoping to find some important influence of philosophy in its coming to be a scientific discipline?

There seems, moreover, to be an unclarity or ambiguity in what we are even looking for on this narrative. Are we looking for some development of age-old concepts that allows philosophy at a given time to serve then as a birthplace of a new science, or are we looking for new topics and methods inside philosophy itself? The suggestion is often the former—philosophers talked long and hard enough about matter that by the seventeenth century a new science of matter could be formed. But, whatever the plausibility of such a story, it is often enough innovations in philosophy that are cited in origin stories of new disciplines. Thus, for example, late nineteenth-century sociology surely was engaged in philosophical debates as it tried to establish its scientific bona fides. But, so it seems to me, the largest philosophical influence here was not hundreds or thousands of years of philosophical discussion of social structure—discussions one can, of course, find—but rather some very specific and at the time quite recent developments in the logic or philosophy of science that were marshaled as resources by theorists such as Max Weber.[6]

In addition to these questions of how to tell such an origin story for any specific discipline, there are larger conceptual issues. If this is not just a story of scientific disciplines arising out of some unspecified nonscientific entity—out of chaos or confusion—we need some general sense of how specifically philosophy could churn out disciplines at all. Why would empirical science begin in philosophy? Moreover, philosophy on this view seems to become ever more empty and, eventually, simply barren. Does philosophy have the ability endlessly to produce more and more sciences? Or have all the interesting and empirically controllable concepts already left the nest? The ever-fruitful because ever-youthful mother of all else is a hard role to fulfill—what evidence do we need to see it in philosophy?

As it happens, whatever the attraction of this narrative structure, very little history of philosophy or history of science is written with these questions in mind. In the history of philosophy that dominates philosophical sensibility—especially in the historical scholarship that enters the curriculum—philosophia perennis is far more prominent. Not every philosopher everywhere and everywhen has written about laws of motion or infinitesimals—thus, when Leibniz speaks of those things, we can skip those sections of his work; those sections are for the historians of science. Leibniz does, however, have a theory of truth, an account of logic, a metaphysics, a doctrine of free will, an account of identity, and so forth. Such topics are prominent in the discussions of Leibniz in the history of philosophy. Newton has a few inchoate thoughts on such matters, which historians of philosophy read on sufferance, allowing historians of science to do the heavy lifting regarding Newton's significance in the history of thought.

While philosophia perennis is clearly a generally conservative account of philosophy, novelty is still appreciated. Indeed, the standard answer to the question "whose thoughts on the eternal questions of philosophy are worth reading?" gives the nod to those who either have novel theories or novel arguments for old theories. Thus, for example, political order and social justice are eternal issues for philosophy but among the early moderns we read Hobbes because his views of political order were particularly novel, well argued, and (thus) influential. Hobbes, on the other hand, is not very good on epistemology, welding a weird mechanical account of mind to an outdated account of geometrical knowledge and a bizarre understanding of names, so Hobbes' epistemology can be passed over, in favor of, say, Locke's or, if one is feeling very bold, Bacon's.

Philosophia perennis does present a sort of frame that makes some sense of the place of history in philosophical discourse. For example, the questions of epistemology can be presented within a framework in which various options for understanding knowledge have been importantly offered by Plato, Aristotle, Descartes, Leibniz, Locke, Hume, Kant, and perhaps a few others. This sets various structures in place: arguments regarding the place of reason and experience, for example, become central, as do issues of structure (foundations versus coherence) among items of knowledge. An entire discourse of epistemology can be built on the basis of considering the epistemological aspects of the work of these figures. Examples can be multiplied: one can trace aesthetic theories through Plato, Aristotle, Hume, and Kant, augmenting these figures with some work by other philosophers who are not prominent in other areas of philosophy, Burke, perhaps, or Shaftesbury. This surely is how much of the history of philosophy is presented in courses and textbooks.

From an historian's perspective, this pedagogical use of philosophia perennis might end up looking very odd, however. After all, what one ultimately achieves here is a sort of unhistorical map of the conceptual terrain of philosophy. The historical philosophers are brought in only to bring out the structure of that map. But, of course, this is as it should be, since the basic idea behind philosophia perennis is exactly antihistorical. The only sort of intellectual development for philosophy that is possible here is precisely the further delineation of the conceptual topography. Since novelty is how a specific philosopher gets into the canon, there is a minor historical element—we read Descartes because he is novel in important ways in his philosophy, for example—but history of philosophy is simply a perspicuous way of developing conceptual maps that one could develop by alternative means (pure conceptual analysis, for example).[7]

From the historical point of view, the problems here are legion. If we are not to beg questions, we need to become quite precise in our understanding of theoretical or argumentative novelty. Descartes is only novel relative to his predecessors and peers, of course, and we have to hope that they were aware of all the previous scholarly work in philosophy so that all the explored options remain in view. So, the whole process begins with someone taken to be novel and important in his own time and then entering the canon. It could turn out that from the distance of a few centuries, the early view of the person involved could strike philosophers as wrong. This is why one genre of historical scholarship in philosophy is the rehabilitation—as when someone argues that Malebranche or Reid did have a viable and novel understanding of truth or knowledge. Moreover, assessments of argument seem to differ across the centuries. Spinoza was viewed throughout much of the eighteenth century as a marginal figure, but he became important for the German Enlightenment, was assessed in a different way by Goethe and the German Romantics, fell into minor status again until he was brought back to prominence in the mid-twentieth century. Can philosophia perennis explain any of this difference in assessment?

For the working historian, the issues run far deeper than this. Two sorts of issues seem to be importantly beyond the scope of any narrative built upon the presuppositions of philosophia perennis. The first is the existence of large swathes of philosophy in various times and places that seem either unconcerned with the "perennial questions" of philosophy or not to fit into the topographic maps offered in the textbooks. Some philosophical work seems very time bound. For example, a fair bit of eighteenth-century philosophy dealt with specific conceptual issues regarding the application of the calculus to reality. While much of this work could be placed within the frame of general issues of atomism versus plenism or the conceptual quandaries of the infinitely small or of the application of mathematics to reality, some of it was deeply connected to specific features of the physics of Newton, the mathematics of analysis, or framed by the specifics of Kant's account of intensive magnitude. Or, again, some of the most influential work in nineteenth-century epistemology seems to be concerned with issues and to make philosophical moves that are much more opaque to current epistemology than was the work of their seventeenth- and eighteenth-century predecessors. For example, there arose in the German epistemological context of the late nineteenth century a rather broad concern with the origin (*Ursprung*) of knowledge, especially among those interested in the foundations of mathematics in a broadly Kantian tradition (Hermann Cohen, Paul Natorp, Edmund Husserl). Such work is largely forgotten and no decent account of Cohen's logic of origin has been offered in a hundred years. While one might wish to make an argument that nineteenth-century epistemology was, thus, all a big mistake, such an account seems to give up on historical understanding rather than to provide it. Why, after all, would history have such periods of deeply mistaken philosophy? Why, especially, would the nineteenth century be such a period, when the historical awareness of philosophers in the nineteenth century seems to have been much higher than that of their early modern predecessors? (Indeed, part of the issue at stake in this work on the origin of knowledge had to do with attempts to theorize the relation of the history of knowledge to the philosophy of knowledge and was informed by much new scholarship at the time in ancient science and philosophy.)

Perhaps an example would render the point more salient. Here is a fairly typical passage from a sort of late neo-Kantian project in epistemology; it is from Bruno Bauch's (1923) *Wahrheit, Wert, und Wirklichkeit* (Truth, Value, and Actuality, 139):

> The logical must, therefore, itself be the archetypical, the objective; the structure of judgment must be the structure of objects; this [structure] cannot be merely projected into that [structure]. And because the structure of judgment is itself the structure of objects and this is, for its part, also the structure of judgment, only thereby can the sphere of possession be projected into subjectivity, can subjectivity possess objectivity at all, can, moreover, objectivity be discovered from subjectivity.

There are very few current epistemologists who could even locate such a view as a possible view in epistemology, let alone explain or elucidate it. Bauch is here working within an articulation of Kantian transcendental logic and attempting to argue that objects must be structured, if they can be known as objects at all, by the very same structure as judgment (the vehicles of knowledge). This structure of judgment is logical structure—thus the world has a logical form. Moreover, he wants to argue that this very fact is the key both to how the mind can know the world (its own structure is the very structure of the world) and how the mind can itself be known as an object in the world.

Bauch is just one of a spectrum of highly influential German epistemologists who took their task to be to articulate versions of key Kantian insights into knowledge (and value) in the context of nineteenth- and twentieth-century scientific advancement. Understanding the work of these thinkers is crucial for understanding the development of positivism and logical empiricism in Germany. (Bauch was Rudolf Carnap's dissertation director.) Yet, the allegedly unhistorical conceptual maps of philosophia perennis provide very little help in understanding what Bauch even took the problem of knowledge to be. Not unrelatedly, the neo-Kantians have almost entirely disappeared from the historical figures canonically read to create that map.

Philosophia perennis, that is to say, ignores a great deal of the detail of the development of philosophy in order to create easily recognizable maps of an allegedly universal terrain.

This is, of course, finally to say that the allegedly unhistorical, eternal map of concepts is very much the product of current concerns. In philosophy, "the eternal" almost always means now. This means that projects that were self-consciously argued against and taken to be refuted by those figures most immediately important for creating current conditions for philosophical study are no longer taken to be in the history of philosophy.[8] This is one reason why even the main themes of nineteenth-century philosophy remain to this day largely forgotten. This doesn't much bother the philosophia perennis advocates precisely because their overall account of philosophy in essence gives history of philosophy nothing to *explain*. History of philosophy is, on this view, reportage, not explanation. Philosophy is intrinsically unhistorical; any historical development is an accident; nothing hangs on there being a trajectory to philosophical concern.[9]

Much of history of philosophy as taught does have the feel of reporting or of history as conceived in secondary school. Students learn that in the seventeenth century Descartes had a position on the mind-body problem and that it is called "dualism." He has some arguments in its favor but he acknowledges some problems and we can think of several more. Students can bring this knowledge of Cartesian dualism with them to systematic courses in metaphysics and philosophy of mind and know what the professor is talking about when she says that no one is a Cartesian dualist anymore. The same goes for hyperbolic doubt, the cogito argument, and Cartesian foundationalism in epistemology. Nothing here hangs on Descartes' situation in the seventeenth century; nothing hangs on his scientific ambitions or achievements; nothing hangs on his theological commitments. Descartes as historical figure does not matter in the history of philosophy, because he does not matter in philosophy. It is only *Cartesian* dualism due to the accident of history that Descartes said it first or, when annoying historians reveal that he did not say it first, best (or really because he said it in the material we still read).

This antihistorical understanding of philosophy does run up against a third sort of narrative, one more explicit within the standard history sequence—especially modern history sequence—than in the relations of that sequence to "systematic" philosophy. It is not unusual for the modern history sequence to be presented as a sequence that goes from rationalism (in seventeenth-century philosophy), through empiricism (in eighteenth-century philosophy), through Kant. After Kant history ends and about seven decades later contemporary philosophy begins. This sequence is often presented in a way that suggests progress: nearly everyone teaches that rationalism gives way to empiricism; some also teach to suggest that empiricism is corrected of both its dogmatism and its tendency to skepticism in Kant. Kant is obviously here playing the role he presented for himself—a synthesis of rationalism and empiricism. In this sequence of the history of philosophy, Hegel plays the guiding, if invisible, role of providing the rationale for the narrative: thesis, antithesis, synthesis.

This narrative structure informs some very general features of the philosophical landscape. It is a poor undergraduate who thinks Leibnizean epistemology is still a viable option or that seventeenth-century understandings of substance or causality survive past Hume and Kant. The modern sequence lines up with an understanding that in ancient philosophy Aristotle's clearheaded empiricism and naturalism supersede the rationalism, idealism, and mysticism of Plato. Throughout, hard-headed empiricism is shown to be superior to more rationalistic rivals. In the versions in which Kant's role as synthesizer and critic are emphasized, one also gets a nod in the direction of a rational foundation for mathematics and logic and, in a different way, a rational foundation for ethics. This solves the problem of what to do with disciplines that do not seem to rely on experience in the way empirical knowledge does—either by being based in proof or in being normative.

From the point of view of real historical development, of course, this narrative is a caricature or farce. It requires that important versions of seventeenth-century empiricism such as Bacon's be submerged as either not really being epistemology (being methodology, say) or relegated to precursor status. It hides important convergences of view on epistemological and metaphysical matters between some who, in order to fit the story, must be forced to be rationalists and those who are made to be empiricists. Influential schools of eighteenth-century rationalism—especially around Christian Wolff—are dismissed as anachronisms. Any historian who had to develop an account of a newly discovered figure in this time period by forcing that figure into the Procrustean bed of a trajectory from rationalist metaphysics to empiricist natural philosophy would find the task unilluminating and a sort of Whiggism of the nineteenth century. Indeed, the overall framework is that of a progression of abstractions—empiricism defeats rationalism—and the sort of detailed empirical work uncovering the commitments of an individual author are submerged in the larger story. Here we definitively see the influence of Hegel—the history of reason supersedes the merely contingent history of what anyone really thought or wrote.[10]

A final narrative for the history of philosophy that must be mentioned—one that is antithetical to philosophia perennis—is a narrative of overcoming or revolution. It is certainly true that in various places and times there have been philosophers who have sought to transcend or overcome or leave behind the history of their subject and start on a new path. Such figures have all the problems of revolutionaries generally—it is very hard to make out how someone can leave behind the entire history of philosophy and still be a philosopher. Perhaps it is not impossible, if you can somehow make some case for continuity of concern but revolution in method, for example—as the early analytic philosophers very often tried to do. Revolutionaries in philosophy additionally have the disadvantage of attempting to revolutionize a discipline with such conservative and inchoate alternative historical narratives. Faced with a proclaimed revolutionary break, a philosophia perennis advocate can always find some way in which the allegedly revolutionary work is either continuous with previous work in philosophy or is simply not philosophy at all. Even the advocate of philosophy as mother discipline can see the revolution as a spin-off of a new discipline, not as a revolution within philosophy.

One problem with the revolutionary narrative from the point of view of history of philosophy is that it often becomes a matter of advocacy or of debunking. That is, histories of logical empiricism may notice the revolutionary intent of the project and then either argue for a genuine revolution or explain how this intent was mistaken or that, whatever the intent, philosophy goes on pretty much as before. Because history of philosophy is often, as we noted early on, undertaken by philosophers, the project of advocacy or rejection often is presented in conceptual terms much like the original arguments. This is to say that the historical work does not often present historical evidence of presence or absence of a revolutionary moment in philosophical history, preferring instead to claim that the project that made that claim was correct or incorrect in its conception of philosophy.

A MORE INTERESTING ROLE FOR HISTORY IN PHILOSOPHY?

Our problem is not to come up with a new, alternate "image of philosophy" to supersede the old image. We have argued that there is no "image of philosophy by which we are possessed"; rather there is a congeries of half-articulated, half-believed, mutually inconsistent images. What the images have in common that is relevant to our current concerns is that they all posit a highly impoverished set of historical questions for the historian of philosophy. These questions are different according to the view in issue but the questions become variations on these: What role did, say, Descartes play in the formation of the modern science of physics? What did Descartes have to say on the perennial issues of philosophy? Did Descartes actually affect a philosophical revolution? Where does Cartesian rationalism stand in the history of epistemology?

Since I am not offering an alternative image for philosophy, I intend to consider only the following question: how would our understanding of philosophy change if we asked better historical questions than these, in particular, if we treated philosophy the way historians of science or intellectual historians treat their fields? How would our understanding of philosophy change if we treated philosophy as a cultural project, undertaken by actual human beings living actual human lives in cultural, political, and intellectual circumstances, circumstances that have varied greatly across the two and a half millennia of Western philosophical history? Here are some expectable results.

History of philosophy would become less individualistic. Instead of focusing enormous attention on a portion of the output of, say, Descartes or Leibniz, it would become more interested in the life course of their ideas. What motivated and occasioned their characteristic philosophical concerns and theses? Who took up the project and how did the project change when it became, say, French Cartesianism or Leibniz-Wolffian philosophy? What role have the figures of Descartes and Leibniz played in the telling of developmental stories of philosophy, logic, physics, or in the development of the intellectual life of France or the Netherlands or Germany? We might come to consider whether it was not the stove-heated room that Descartes retreated to in 1619 that was important as much as it was that the stove-heated room was in Bavaria and Descartes was there because he was with the Dutch army for the crowning of the Holy Roman Emperor, an important occasioning cause of the Thirty Years War. That is, we might come to understand more about the content of philosophical doctrines by embedding those who articulated them in their rich historical contexts.

Perforce, in becoming less individualist, such history will become less focused on "the great philosophers." Leibniz lived on throughout the eighteenth century in Germany in large measure not so much due to his own individual genius as due to generations of philosophers who kept alive a project that they saw as importantly continuous with Leibniz's own. Christian Wolff is only the most prominent of these philosophers. More attention would be given to knowing exactly what, for example, Kant learned and taught in Königsberg as he felt moved to change his allegiance from such metaphysical philosophy to his own critical philosophy. Once the hold of the ranking system of philosophers (Kant higher than Mendelsohn, Mendelsohn higher than Wolff) abates, it will no longer seem an odd waste of time for someone to investigate in detail, for example, the nineteenth-century German idealist logicians—and to do so in order to uncover the details of their own work and not just to answer questions about their relations to Gottlob Frege. Nor will the project of interpretation of a key text be seen as accomplished when that text is put into conversation with texts written scores or hundreds of years before or after it. Carnap's *Aufbau* will be understood in its context, not in its vague relations to Hume or Locke—unless someone finally brings forth evidence that Carnap took himself to be pushing forward a Humean or Lockean project or to be commenting on their views. Philosophical attention will then range more broadly within the writings of any philosopher and more broadly into the philosophical and other intellectual contexts in which that philosopher wrote.

Philosophy will become less confident that it has figured out which issues matter to it and which fields belong to it. It could be possible to find out that the organization of philosophy into fields was significantly different in Paris in 1800 than it was in Los Angeles in 1970. It might be possible to discover that what was at stake in philosophy in the life of ancient Greek culture is only marginally related to what is at stake culturally in philosophy in Berlin in 1870 (while attending to the ways in which ancient Greek questions about the place of philosophy were discussed and relied upon in Berlin). Attention to the place of philosophy relative to other cultural projects—science, literature, and religion, for example—might lead historians of philosophy to ask different questions. Rather than being puzzled about why Kant didn't think ethical life was for the promotion of happiness, we might become puzzled about how nineteenth-century figures have been so persuasive within philosophy in arguing that ethical life was about promoting happiness. After all, the Protestant religious traditions informing much of early modern and Enlightenment moral theory—and certainly Kant's own moral theory—surely did not suggest that the good life was the happy life.

Philosophy could also begin to invite curiosity about topics typically entirely left out of account. We have almost no histories of philosophical curricula or textbooks despite the fact that the undergraduate classroom is where nearly all professional philosophers over the past two hundred years have begun their training and imbibed their sense of the overall structure of the field—and it is also where the vast majority of today's philosophers earn their pay. Moreover, textbook traditions have been interestingly variable in the teaching of philosophy—both across time (Kant taught almost entirely from textbooks; very few current advanced courses use textbooks) and across discipline (the textbook tradition is very robust in logic and barely exists in some other fields; the anthology of great historical works is perhaps the most common current textbook in introduction to philosophy, a fact that becomes very confusing to students who find out later on in their studies how little attention contemporary philosophical research pays to the figures they have been asked to read). We have very few treatments of PhD training, of the formation of journals and societies, or of other professional activities that structure or have structured philosophical research (research funding regimes, program development, etc.). These are all "external" to real philosophy on the narratives that structure philosophers' sense of history; but these activities constitute the conditions of the continuance of philosophy according to the standards of intellectual historians or historians of science.

Philosophy could also profit from attention to different sorts of questions—instead of asking about specific figures and the theses they present, we could become interested in, for example, the figure of the philosopher *tout court*. Arguably the greatest continuing reason for cultural interest in philosophy is due to the figure of the philosopher as the good and wise person. This sort of philosopher is a far cry from the professional philosopher in today's academic philosophy. More than that, the image of the philosopher as the profound genius or the exemplary human being was explicitly attacked at the turn of the twentieth century as having been detrimental to the development of the field of philosophy. The alternative images for the philosopher on offer—mainly modeled on the research scientist—arguably did not take hold of the imagination of learned people outside of philosophy, and this fact accounts for much of the trouble analytic philosophy in particular has in convincing those outside that it even is philosophy. If analytic philosophy has never controlled the image of the philosopher, it stands in clear and present danger of losing control of what counts as philosophy. Arguments, often more puffed-up finger wagging or frustrated screeds, offered by analytic philosophers that, say, Nietzsche or Heidegger or Derrida was not a philosopher at all are predicated on a vision of what it is to be a philosopher that virtually no one outside of analytic philosophy shares.

PHILOSOPHY AND HISTORY OF PHILOSOPHY SO CONCEIVED

It might be allowed by critics that philosophy would change under such a view of what it is to do history of philosophy. But, they would continue, philosophy would be worse for that. Instead of inquiring after Descartes' theory of mind, we would be fussing over what textbooks he learned from. Rather than concentrate on the lasting significance of Kant's transcendental turn, we would be immersed in detailed discussions of his relationships with contemporaries. In general, the substance of philosophy would get lost in an externalist contextualism.

This is untrue for at least two important reasons. First, many of the detailed and deeply interesting philosophical positions even the historically great philosophers enunciated were taken up within the context of genuine arguments with contemporaries or due to their historically conditioned sense of what was central to philosophy. Contextualism is simply a more detailed and sympathetic working through of philosophical argumentation. It does concentrate on the significance the historical actors in the dispute thought the dispute had rather than on our subsequent reflections on it. But it does not deny and even takes up as an important historical theme the fact that philosophers often do take up arguments, methods, and positions from distant historical antecedents. Second, the internal-external distinction is predicated on *our* sense of what is and is not philosophy, and it is precisely this sense that is most importantly historically thematized in the sort of contextualism I am considering. Leibniz absolutely did not think his work in the calculus was unrelated to his work in metaphysics—knowing in detail Leibniz's work in the calculus can enrich our sense of his philosophical projects. It is contemporary philosophical self-understanding that dictates that complicated mathematical developments are not relevant to metaphysics—and this renders most of Leibniz's readers technically incompetent in central areas of his philosophical interests. Exploring the richness of historically available conceptions of philosophy can, in turn, enrich our own sense of what is and what is not a proper philosophical project.

It is true that the history of philosophy I envision here shares tone-lowering gestures with post-Kuhnian history of science: we concentrate on persons, organizations, schools in philosophy, on what they were actually doing and what they were actually motivated by; history of philosophy becomes interested in the contingent, the bounded, and human.[11] Such tone-lowering gestures in the case of philosophy would, ironically, remind us that in various places and times philosophy has had important and complicated relations with religion, science, art, and other important human activities and endeavors. In some places and some times philosophy has really mattered to human culture. A disillusioned history of philosophy of the sort I am imagining need not be a demoralized one—indeed, it might allow more seriousness to creep back into the philosophical project than one routinely sees among those who exploit their position as heirs of Socrates, Descartes, or Kant even as they resolutely avoid any serious engagement with the work of such figures. The envisioned history of philosophy places the scope and significance of philosophy itself firmly in view—and this topic is worth occasional reflection by even the narrowest specialist in philosophy. Such a specialist, after all, relies on a widespread cultural sense of the significance of philosophical work to set the conditions under which he can engage in his own projects, even if he neither shares the view himself nor engages in projects of cultural or philosophical significance.

Kuhn introduced his new image of science in self-conscious opposition to "the image of science by which we are possessed" and he did so on historiographic grounds. He knew that the image of the history of science contained within the received image of science served scientific, not historical, purposes. Kuhn did not wish to disrupt the practices of working scientists and I very much doubt he would have expected the accounts of history given as exemplary vignettes in science textbooks to change to match the account of the history of science provided in SSR. Indeed, for a while one question raised in the reaction to Kuhn was whether scientists should study the history of science as offered by Kuhn—given that this history could not discharge the scientific function that history of science had on the received view. One consequence of Kuhn's intervention, then, was further to distinguish and fix the distinction between real working historians of science and scientists.

Suppose, then, we implemented what I have argued is more pertinent historical inquiry in philosophy. Would this further distinguish the roles of historian of philosophy and philosopher? Would it, thus, remove history of philosophy even further from the business of the standard working philosopher? Perhaps, but I doubt it. Here are some considerations that tell in the other direction. First, it is almost impossible for contemporary philosophers to place their views into the philosophical landscapes of their own subdisciplines without making reference to historically given views. Thus, phrases such as "Cartesian dualism," "Spinozan monism," "Kantian critique," or "Moorean intuitionism" are unlikely to disappear from philosophical vocabulary in use. This being the case, history courses will need to be part of the training of philosophers—on pain of these terms losing all traction. (It is true that terms like "Newtonian physics" don't induce physicists-in-training to rush off and read Newton, but this is because Newtonian physics is not a marker of a possible position in a current controversy. On the other hand, Cartesian dualism is part of the very framing of the mind-body problem itself.) History courses done with real historical content of the sort I am envisioning, then, would provide philosophers-in-training a richer, more nuanced view not only of the positions of the philosophers they are reading but of the whole question of what makes the issues they engage with philosophical issues (in the work of the historical philosophers themselves).

The ultimate argument that any working philosopher can provide that her work is in fact philosophy is an (implicit, in the normal case) argument that there is some historical tradition or trajectory that renders the questions she asks and the answers she gives philosophical. Thus, the final ground of all philosophical work is the history of philosophy. A richer and more empirically grounded sense of what historical traditions in philosophy actually are and what various options have been in the prosecution of philosophical work, far from disconnecting philosophy from the richer history of it that I have argued for, could affect a revolution in the self-understanding of the working philosopher and yield a richer, more satisfying set of philosophical projects for the twenty-first century.

NOTES

1. This essay has been substantially improved by the close reading and subtle remarks of the editors, for which I am grateful.
2. For Kuhn's influence on sociology of science, see, for example, Barnes and Dolby (1970); for Kuhn's doubts about the sociology of science that invoked his name, see Kuhn (2000, especially Chapter 5). For Kuhn and rhetoric of science, see Harris (1997). For Popper's early reaction see Popper (1970) and indeed the entire Lakatos and Musgrave volume (1970). For Rorty's use of Kuhn see Rorty (1979). For Kuhn's post-SSR engagement with philosophy of science, see Kuhn (2000).
3. Of course, Kuhn did have an interest in philosophy dating back to his undergraduate days—indeed that interest, sparked by his understanding of Kant, ultimately played a large role in how he framed his own account of science. See Richardson (2007).
4. There is an irony here. Kuhn's account of science rendered it *more* isolated from the rest of learned culture than it was in visions of science as perpetual revolution in the work of philosophers like Popper. Popper saw Kuhn's normal scientists as dogmatists, *Fachidioten*. It was Kuhn's image of science reaching for conceptual resources outside of itself in moments of crisis that fired the imagination of his readers. Kuhn himself thought he was mainly doing "internal history."
5. Popper saw this quite clearly; his objection was to the authoritarian structure of normal, not revolutionary, science. On Kuhn's influence on conservative theories of science, see Shapin (1995).
6. This seems to have been Weber's view (see Weber [1904] 1949); in any case, the new logic of the human sciences allowed the scientific status of sociology to be achieved.

7. Of course, once the canonical figures are set and it is widely presumed that Descartes, for example, is novel in his treatment of the "perennial questions" of philosophy, historical work can be done to see if his work was in fact novel in that way. Often such scholarship is treated as "mere history" by those who read Descartes for what he says about, say, knowledge and don't care whether the implicit historical claim that he said it first or best is right.

8. This extraordinary measure is rarely undertaken explicitly, but see Roger Scruton's extrusion of Hermann Lotze from the history of philosophy in the introduction to his *Modern Philosophy* (2004, xiv–xv). Scruton offers a methodological version of philosophia perennis—philosophy has certain features (abstraction, concern for argument and truth) that are recognizable across all time; those not properly within the history of philosophy, like Lotze and Foucault, can find a place in the history of ideas. The weary, frustrated reader of Scruton's tome cannot help but think that what he excludes must be more interesting than what he leaves in the history of philosophy.

9. Vasso Kindi has reminded me that in certain historiographical traditions that include contemporary historians of various stripes, it is denied that historical work does explain. There are complicated issues here of what explanation might mean in the work of history, but in any case real historical work does more than illustrate an already given conceptual terrain.

10. When Lakatos wrote that in history of science what really happened could be confined to the footnotes, he was bringing sensibilities from a particular brand of history of philosophy into history of science. Lakatos, of course, knew that, having been raised on Hegel; most analytic historians of philosophy do not know that their work is given shape and meaning by Hegelian sensibilities in history of philosophy.

11. On tone-lowering as a theme in recent history of science, see Shapin (2010, Chapter 1). Since many of Shapin's historical actors are natural philosophers, he has been for decades engaged in the sort of work I am advocating. Applying the resources of science and technology studiers to history of philosophy in ways continuous with what I am arguing has happened many times. A recent effort to both apply and thematize the application of methods from science and technology studies in doing history of philosophy is Francesca Bordogna's recent assay in "philosophy studies" (Bordogna 2008).

REFERENCES

Barnes, Barry, and R. G. A. Dolby. 1970. "Scientific Ethos: A Deviant Point of View." *European Journal of Sociology* 11: 3–25.

Bauch, Bruno. 1923. *Wahrheit, Wert, und Wirklichkeit*. Leipzig: F. Meiner.

Bordogna, Francesca. 2008. *William James at the Boundaries: Philosophy, Science, and the Geography of Knowledge*. Chicago: The University of Chicago Press.

Harris, Randy Allen. 1997. "Introduction." *Landmark Essays on Rhetoric of Science*. Mahwh, NJ: Lawrence Erlbaum, xi–xl.

Kuhn, Thomas S. 2000. *The Road Since Structure*. Chicago: University of Chicago Press.

Kuklick, Bruce. 1984. "Seven Thinkers and How They Grew: Descartes, Spinoza, Leibniz; Locke, Berkeley, Hume; Kant." In Richard Rorty, Jerome B. Schneewind, and Quentin Skinner (eds.), *Philosophy in History*. Cambridge: Cambridge University Press, 125–139.

Lakatos, Imre. 1978. "History of Science and Its Rational Reconstructions." In *Methodology of Scientific Research Programmes*. Cambridge: Cambridge University Press, 102–138.

Lakatos, Imre, and Alan Musgrave, eds. 1970. *Criticism and the Growth of Knowledge*. Cambridge: Cambridge University Press.

Popper, Karl. "Normal Science and Its Dangers." In Lakatos and Musgrave (1970), 51–58.

Richardson, Alan W. 2007. "'That Sort of Everyday Image of Logical Positivism': Thomas Kuhn and the Decline of Logical Empiricist Philosophy of Science." In. A. W. Richardson and T. E. Uebel (eds.), *The Cambridge Companion to Logical Empiricism*. Cambridge: Cambridge University Press, 346–369.

Rorty, Richard. 1979. *Philosophy and the Mirror of Nature*. Princeton, NJ: Princeton University Press.

Scruton, Roger. 2004. *Modern Philosophy*. 2nd ed. London: Random Thoughts UK.

Shapin, Steven. 1995. "Here and Everywhere: Sociology of Scientific Knowledge." *Annual Review of Sociology* 21: 289–321.

Shapin, Steven. 2010. *Never Pure*. Baltimore: The Johns Hopkins University Press.

Weber, Max. 1949. "Objectivity in Social Science and Social Policy." In Edward A. Shils and Henry A. Finch (eds.), *Max Weber on the Methodology of the Social Sciences*. Glencoe, IL: The Free Press, 49–112. Original published in 1904.

Contributors

Hanne Andersen is Associate Professor at the Center for Science Studies, Aarhus University. Among her publications are *On Kuhn* (Wadsworth 2001) and *The Cognitive Structure of Scientific Revolutions* (coauthored with Peter Barker and Xiang Chen, Cambridge University Press, 2006). Among her research interests are incommensurability and the realism debate, the nature of interdisciplinarity, and social epistemology.

Theodore Arabatzis is Associate Professor in the Department of Philosophy and History of Science at the University of Athens, Greece. He has published on the history of late nineteenth- and early twentieth-century physical sciences, and on philosophical and historiographical issues concerning conceptual change, scientific realism, and experimentation. He is the author of *Representing Electrons: A Biographical Approach to Theoretical Entities* (Chicago, 2006) and co-editor of *Metascience*.

Alexander Bird is Professor of Philosophy at the University of Bristol, having previously been Reader and Head of the Department of Philosophy at the University of Edinburgh. His work is in the metaphysics and epistemology of science and medicine, and he has a special interest in the philosophy of Thomas Kuhn. His study of Kuhn's work, *Thomas Kuhn*, was published by Acumen and Princeton in 2000. He has also published *Philosophy of Science* (Routledge, 1998) and *Nature's Metaphysics* (Oxford, 2007).

Hasok Chang is Hans Rausing Professor of History and Philosophy of Science at the University of Cambridge. He received his degrees from Caltech and Stanford, and taught at University College, London, from 1995 to 2010. His main research area is the history and philosophy of the physical sciences from the eighteenth century onward. He is the author of *Inventing Temperature: Measurement and Scientific Progress* (Oxford, 2004) and *Is Water H_2O? Evidence, Realism and Pluralism* (Springer, 2012).

Gürol Irzik is Professor of Philosophy at Sabanci University, Istanbul. He received his PhD from the History and Philosophy of Science Department at Indiana University, Bloomington, in 1986. He has published on causal modeling, Carnap, Popper, Kuhn, and science education. Currently, he is interested in the commercialization of science and in political philosophy of science.

Vasso Kindi is Associate Professor in the Department of Philosophy and History of Science at the University of Athens, Greece. She is the author of *Kuhn and Wittgenstein: Philosophical Investigation of the Structure of Scientific Revolutions* (Smili, 1995, in Greek). She has published on Kuhn, Wittgenstein, conceptual change, philosophy of history, and ethics.

Jouni-Matti Kuukkanen is a postdoctoral research fellow in the project *Philosophical Foundations of the Historiography of Science*, based at Leiden University. He wrote his doctoral dissertation on Thomas Kuhn at the University of Edinburgh, and it was subsequently published as *Meaning Changes: A Study of Thomas Kuhn's Philosophy* (VDM Verlag Dr. Muller, 2008). Kuukkanen is currently preparing a book on postnarrativist philosophy of historiography.

James A. Marcum is Professor of Philosophy, a member of the Institute for Biomedical Studies, and director of the Medical Humanities Program, at Baylor University. He earned doctorates in philosophy from Boston College and in physiology from the University of Cincinnati Medical College. He is the author of *Thomas Kuhn's Revolution: An Historical Philosophy of Science* (Studies in American Philosophy Series, Continuum, 2008).

Thomas Nickles is Foundation Professor of Philosophy, University of Nevada, Reno, United States; external faculty member, University of Catania, Italy; and member of the PratiSciens group, Archives Poincaré, Université Nancy 2, France. A historically oriented philosopher of science, his central interests are scientific and technological innovation, problem solving, and scientific change. He is editor of *Thomas Kuhn* (Cambridge University Press, 2003) and coeditor of *Characterizing the Robustness of Science* (Springer, 2012).

Rupert Read is Reader in Philosophy at the University of East Anglia. He authored, coauthored, or coedited the following books: *Kuhn: Philosopher of Scientific Revolution* (Polity, 2002), *The New Wittgenstein* (Routledge, 2000), *The New Hume Debate* (Routledge, 2000), *Film as Philosophy: Essays on Cinema after Wittgenstein and Cavell* (Palgrave, 2005), *Applying Wittgenstein* (Continuum, 2007), *Philosophy for Life*

(Continuum, 2007), and *There Is No Such Thing as a Social Science* (Ashgate, 2008). His next work (cowritten with Phil Hutchinson) will be a monograph on Wittgenstein's method.

Alan Richardson is Professor of Philosophy and Chair of the Science and Technology Studies Graduate Program at the University of British Columbia. A former President of the International Society for the History of Philosophy of Science (HOPOS), he works mainly on topics in the history of philosophy of science in the twentieth century, especially the history of logical empiricism.

Wes Sharrock is Professor of Sociology at the University of Manchester. Among his books are *Kuhn: Philosopher of Scientific Revolution* (coauthored with Rupert Read, Polity, 2002) and *Garfinkel and Ethnomethodology* (coedited with Michael Lynch, Sage, 2003). He has published on ethnomethodology, relativism, and rules.

Index

/